Designing and Building a Grandfather Clock

Designing and Building a Grandfather Clock

GARY WILLIAMS

Illustrations by the author

San Diego • New York
A. S. Barnes & Company, Inc.
In London:
The Tantivy Press

Designing and Building a Grandfather Clock
text copyright © 1980 by
A.S. Barnes and Co., Inc.

The Tantivy Press
Magdalen House
136-148 Tooley Street
London, SE1 2TT, England

All rights reserved under International and Pan American Copyright Conventions.
No part of this book may be reproduced in any manner whatsoever without written permission from the publisher, except in the case of brief quotations embodied in reviews and articles.

First Edition
Manufactured in the United States of America
For information write to A.S. Barnes and Company, Inc.,
P.O. Box 3051, San Diego, CA 92038

Library of Congress Cataloging In Publication Data
Williams, Gary, 1945-
   Designing and building a grandfather clock.

   1. Clock and watch making. 2. Longcase clocks.
I. Title.
TS545.W53   681'.113   79-24335
ISBN 0-498-02209-9

Manufactured in the United States of America

for Lee

CONTENTS

Preface   11

Acknowledgments   13

1 **Quality**   17

2 **Tools**   20
    One: Keeping Your Fingers   20
    Two: The Shop   20
        Benches   20
        Clean Room   21
        Pads   21
        Gluing Jig   21
    Three: The Equipment   21
    Four: Spending Your Money   22
    Five: Adjusting Power Tools   24
    Six: Sharpening Without Stones   24
        Slick-ems   29

3 **The Movement**   30
    One: Choosing   30
        Rod Chimes   30
        Tube Chimes   31
        The Difference   31
        Other Movements   31
        Silencing   32
    Two: Mounting The Movement   32
        Mounting the Dial   34
        Cutting A Round Hole   36
    Three: Adjusting   37
        Crutch   37
        Pendulum Bob   37
        Suspension Spring   37
        Chime Synchronizing   38
        Minute Hand Correction   38
        The Handshaft Nut   39
        Oiling   39

4 Designing   40
    One: The Fixed Elements   41
        Crowns   43
    Two: The Narrow-Waisted Case   44
        Designing A Grandfather Clock   46
    Three: Notes on Design   47
        Design Sampler   48

5 Construction   50
    One: Making the Patterns   51
    Two: The Door   53
        Warping   53
        Choosing Grains   55
    Three: The Box   60
        The Bridge   60
        The Front   62
        The Sides   63
        The Back   66
        Straightlining Long Boards   67
    Four: Putting It Together   64
        Squaring Up   70
    Five: The Roof   71
        Shoulder Caps   73
    Six: The Bottom Box   73

6 The Gingerbread   75
    One: Foot Levelers   75
    Two: Diagonal Grains   75
    Three: Large Cove Moldings   77
    Four: Mitered Moldings   78
    Five: Front Moldings   79
    Six: Split Turnings   80
    Seven: Swan's Neck Molding   81
    Eight: Raised Panels   82
        Raised Panels Are Stable   83
        Making the Rails   84
        Radial Arm Setup for Cutting Rail Ends   85
        Cutting Tenons with a Dado Head   86
        Raising the Panels   86
        Table Saw Setup   86
        Radial Arm Setup   87
    Nine: Hinges   88
    Ten: Locks   89
    Eleven: Installing the Movement   91
        The Bracket   91
        Measuring   92

7 The Finish   94
    One: Sanding   94
        Belt Sanding   95
        Orbital Sanding   95
    Two: Polishing   96
    Three: Oiling   97
    Four: Maintaining the Finish   98

8 Leaded Glass   100
    One: Design   101
        Patterns   103
        The Cartoon   104
        The Cutting Patterns   105
    Two: Choosing Glass   107
        Color   107
        Types of Glass   108
    Three: Cutting   109
        Grozing   112
        Cutting the Dial Glass   112
        Deburring   113
    Four: Glazing   113
        Pulling the Lead   115
    Five: Soldering   118
    Six: Cementing   120
        The Recipe   120
        Cleaning   121
    Seven: Installing   121

9 Wood Carving   123
    One: Tools   123
        The Bench   123
        Cutting Tools   123
    Two: The Cuts   125
        Stop Cut   125
        Lowering   127
        The Right (and Wrong) Direction   130
        Modeling   130
        A Simple Flower   131
        The Right Direction (Again)   132
        A Simple Leaf   132
        Grapes   133
        Acorns   134
    Three: Using a Router   135
    Four: Design   136

10 Sources   141
    Clock Movements, Dials, Plans, Kits, and Parts   141
    Goodies   143

PREFACE

The thing is, nobody needs a grandfather clock any more. You probably have one of those obnoxious electric things that buzz, beside your bed or on the stove or someplace . . . or a digital—everything's digital. There's even a digital grandfather clock, if you can imagine that. And who's without a wristwatch? Kids get their first one from Santa Claus when they're about five. Do we need to go out to the hall to check with the Grand Old Hall Clock when we want to know what time it is? Most of us don't even have a hall any more, unless you count that tunnellike affair leading to the bedroom where the thermostat and the linen closet are. And as for chimes, who could hear them over the stereo?

You don't need a hall clock. What you need is a nice table, a nice bookcase, something to put the salt shaker collection in. And when you buy or build things you need, you probably don't get terribly extravagant; you shop for a bargain, and try to keep it reasonably functional; after all, you need lots of things, and there is only so much money.

Toys, on the other hand, are a different matter. When it comes to toys and treasures and art, we have to use a whole new set of rationalizations. We're not dealing with need anymore, we're dealing with want. When we want a thing, just for the sheer joy and pride in having it, we stop thinking about it being functional and start thinking about it being fine. We tell ourselves (and each other) that the toy is an investment, but, if pressed, we'd have to admit it's just good old-fashioned unabashed indulgence.

And where would we be without that kind of indulgence? If nobody ever did anything just for the sake of beauty or quality or pride or fun, what a gray, functional place the world would be.

Build your hall clock because you want it. And build it fine. It will unquestionably be handed down to your children and probably their children as one of the family's most prized possessions. How much more prized if Grandpa built it himself. And how much more yet if it was his own design, out of his head, and his personality; a visible expression, generations hence, of what the old man must have been like.

A grandfather clock is, and should be, an object to be loved. It should be built with love and handed down with love. Cabinetmakers who come into my shop and announce that they have built a hundred or two hundred clocks and got $300 apiece for them don't impress me. They've missed the point. Show me a person who's built one clock with love, and I'm impressed.

ACKNOWLEDGEMENTS

I would like to thank Walter Judson for much help, especially with the stained glass chapter, and Dale Price, Jim Schubert, Mark McGarity, Ken Burchett, Mel Walker, and Damon Vincent for teaching me woodworking.

And I would like to thank Jack Spaulding for teaching me about quality.

Designing
and Building
a Grandfather Clock

1

QUALITY

There is a small college in the Midwest where the students pay no tuition. Instead of paying, they work at various jobs and crafts in shops set up around campus. I went there a few years ago to set up a shop to build grandfather clocks. The students who worked with me had to learn the craft in a relatively short time, and they had to learn it well, because the clocks we built were for sale, and they were expensive.

While they all had some industrial arts and woodworking training, they had little or no experience with hall clocks. Each student worked twenty hours a week in the shop, and in order to make that time productive we developed some techniques for making the joinery as accurate and as foolproof as possible. We had to bend some rules and abandon some time-honored procedures, but what we came up with were fine, tight, well-joined cases, much higher in quality than most commercially available clocks.

Let me emphasize that the techniques we used (and which I still use) were not shortcuts that lessened quality; they made high quality possible for craftsmen with limited experience.

We were in the business of selling clocks, but it was, after all, a college, so we were really more interested in learning something than in just making money. What that meant was that we approached the project more as non-professional craftsmen than as businessmen. While non-professional craftsmen and furniture businessmen are both interested in maximum return on their investments, the essential difference is in the form that return takes. For a businessman, the return is simply money. But since a non-professional craftsman is more than likely building the clock for himself, the return he's interested in is the best clock possible.

The two returns require opposite approaches to the investment. In order to come up with as much money as possible, a businessman has to make sure he spends as little as possible in the first place on materials and labor. His extraordinary machinery can help keep labor costs down, but the only way to keep material costs low is to use cheaper, hence inferior, materials.

It only makes sense that a craftsman, interested in a return measured in quality instead of dollars, would use the finest (hence the most expensive) appropriate materials he could afford.

And to me this applies even when cheaper materials look like the real thing. *You* know that veneered plywood is not walnut, just as you know that "gold-plated" does not mean gold. You also know that over the years and generations gold plating will wear off, and that veneer will become nicked and damaged.

And it applies even to the places which "don't show." In my clocks I put a three-inch thick slab of walnut in the bottom for nothing more than weight.

The wood investment represents about ten to fifteen percent of the value of the finished clock. If you were to use all the plywood you could possibly get away with, you might save yourself five percent of the finished value—significant enough for a profit-motivated business, perhaps, but hardly for a non-professional craftsman. (Actually, plywood saves the factory on labor costs as well as materials, but presumably you and I are in this because we like the labor.)

My students were being trained in industrial education, and I had to turn them around a little. I had to teach them to waste wood, or at least to be extravagant with it. If the grain pattern we wanted for the waist door happened to be smack in the middle of a big board, we'd cut it right out and let the ends fall where they might. Those ends would be used eventually, but even so, that's not a very economical practice. If the return on your investment is to be quality, then an otherwise acceptable board with a small knot in it will be out of the question. Put it aside and find one without a knot. Or go buy another one if necessary. A hall clock deserves the kind of extravagance which causes the craftsman to unblinkingly throw out a finished door when the router slips, or a joint turns out bad. One good putty fill can negate the return in pride and quality on your entire investment.

The joinery techniques in this book are simple enough and foolproof enough to eliminate the problem of bad joints, and a proper attitude about the cost of rough lumber in relation to the value of the finished project should keep you in perspective with knots and minor goofs, so throw away your putty can.

Our two guiding principles will be quality and simplicity. We'll discuss leaded glass, because it enhances quality in the clock case. It also happens to greatly simplify construction of some of the wooden parts, as we shall see in later chapters. We're not going to get into stained glass beyond what you need to know for a clock case.

The same applies to wood carving; there is a chapter on wood carving which deals with the basics of the kind of surface decoration you might want to put on a clock.

The experience of building a clock, and shifting the emphasis from economy and function to quality and beauty, will stay with you the rest of your woodworking days, and may well have an effect upon everything you build from now on.

2
TOOLS

One: Keeping Your Fingers

Fingers are like children; they're part of you and you love them dearly, but if they're hanging out somewhere and you don't know where they are, chances are they're in trouble.

You may not have thought about it, but your right hand very often does NOT know what your left hand is doing, because your brain is concentrating so hard on getting the cut right that it forgets to tell the fingers to get out of the way. One of the two rules I live by in the shop is this: after you're all set up and BEFORE you start feeding the wood to the blade, take inventory of your fingers; know where they all are in relation to the blade and in relation to where the blade is going to be coming OUT of the wood. The second rule is: know in which direction the blade can throw things, and stand in some other direction. Give *any machine in the shop* just half a chance, and it'll make you bleed.

Two: The Shop

Benches

Try to arrange a long, narrow workbench, not much bigger than the case you plan to build, that you can walk all around to get to all sides. It should probably be two or three inches lower than standard bench height (to make working on the assembled case easier) and the top should be as flat and true as possible.

If you're not in the habit of putting everything away as soon as you're finished with it, have two workbenches: one to work on and one to pile junk on. It's difficult and frustrating to try to work on a cluttered workbench.

Clean Room

You should have a "clean room" for the movement to hang in until the case is done; in the house, preferably, if you can keep little fingers out of it. Never have the movement in the shop unless it's absolutely necessary. Clock oil plus sawdust equals crud.

Pads

Make three or four pads to protect sanded wood from the bench surface. Cut two-inch strips about two feet long of carpet, carpet pad, foam rubber, or some such—even corrugated cardboard will do. But you don't need big, cumbersome sheets; two-inch-wide strips do nicely.

Gluing Jig

When you make long, narrow door rails, it's a good idea to laminate them (the job is described in chapter 5). To do it right, you need a true surface to clamp to. Make one as shown out of ¾" plywood.

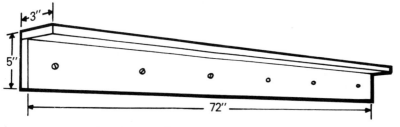

Attach to wall or edge of workbench.

Three: The Equipment

You'll need a table saw, a skillsaw, and a band or saber saw to cut the case out, a router for rabbeting, and a drill, a doweling jig, and lots of clamps to put it together. A radial arm or shaper for making your own moldings would be very desirable, and when you start finishing you'll bless the day you bought (or rented) a belt sander and a GOOD orbital sander. About the best woodworking investment you could make, next to a good square table saw, would be a sander like the Rockwell Commercial Model 4485 orbital. It's 10,000 OPM (most are around 4000), it's direct drive and quiet (most are gear drive and noisy), and it do make the sawdust fly. It doesn't have "dual action" or sawdust pickup or any of that; it's a commercial tool and it almost makes a pleasure out of a job nobody likes. You get what you pay for: you can buy an orbital sander for fifteen dollars; the Rockwell 4485 will cost you about a hundred.

Four: Spending Your Money

If you're going to buy some equipment, allow me to offer the following opinions:
• A radial arm saw is more versatile than a table saw, but it's much more difficult to square up, because adjustments (and they are many) require you to move large masses of steel, and that works against gnat-hair-fine tuning. A table saw has only that little miter gauge which can be pressed directly against a square and tightened in place, and a knob for blade tilt, which you adjust as if you were tuning a radio.
• Spend your money where it counts most: on precision where precision is necessary (table saw, jointer), on durability where that's important (electric drill), on high-quality steel (chisels), and on power for difficult jobs (sanders). On tools which don't require a lot of precision adjustments you can save money.
• A good used bandsaw, for example, should do fine, as long as the blade stays on the wheels and the bearings aren't shot, because a bandsaw is not a precise tool (it's not used for joints).
• Your router should be in good shape, but there's not a lot that can get out of whack with a router, so there's no need to buy the most expensive one.
• Nothing terribly precise about a wood lathe, either; just check for parts that wiggle that shouldn't, and make sure the centers are reasonably well aligned.
• A skillsaw is very handy in a cabinet shop, but it won't be doing contractor duty, so it doesn't have to be anything special.
• A good, many-toothed carbide blade for your bench saw will make you happy, not only because it stays sharp, but because it crosscuts like a planer blade and also rips, which a planer blade won't do, so you aren't changing blades every five minutes.

I don't mean to suggest that you need to buy every piece of equipment there is, but if you're at all serious you ought to have at least some kind of a bench saw for accuracy, and every one of the hand-held power tools you can finagle will save you a lot of hard work.

One general rule of thumb will help you pick tools for durability:

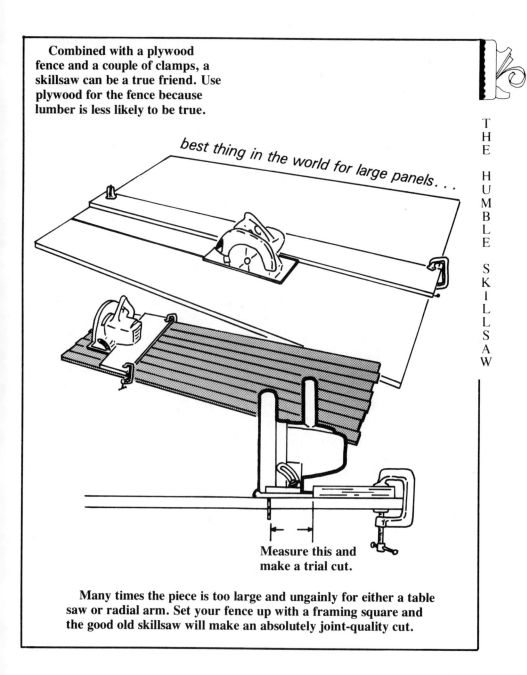

Combined with a plywood fence and a couple of clamps, a skillsaw can be a true friend. Use plywood for the fence because lumber is less likely to be true.

best thing in the world for large panels...

Measure this and make a trial cut.

Many times the piece is too large and ungainly for either a table saw or radial arm. Set your fence up with a framing square and the good old skillsaw will make an absolutely joint-quality cut.

metal against metal equals wear. That means try to get ball bearings instead of sleeve bearings whenever you can, and look for direct drive or belt drive and avoid gear-driven equipment if at all possible. All that noise they make is metal against metal. Turn a machine on; if it screams like a skillsaw, it's gear driven. Pass it up if you can. (Some drill presses and radial arm saws are gear driven.) You don't have much choice with skillsaws and electric hand drills; they almost have to be gear driven.

Tool selection is a common-sense business. Try to be realistic about where you will count on quality most and spend your money there. (And quality doesn't necessarily mean size and power; a table saw needs to have a fair amount of power, but mainly it needs to be sturdy and conveniently adjusted. How many four-by-fours do you cut in the average cabinet shop? An 8- or 9-inch blade will cut ¾" stock as well as a 10- or 12-inch one.)

Five: Adjusting Power Tools

If everything in joinery could be condensed to one basic principle, it would be accuracy of the cut. In the end, whether the thing goes together squarely (or at all) depends directly on how much care was taken when all the sides and ends passed the spinning steel. If you keep the importance of that fact in mind you'll do certain things: you'll always use a sharp pencil or a striking knife, you'll measure from one piece to the other if they have to be the same size, or, better, you'll cut them both in the same pass. You'll realize that the degree markings on the machine are to be ignored and that the only thing you can trust is your square. And, above all, you'll ignore the 90- and 45-degree stops in the machine. They're put there for lazy people, and they're a pain in the neck. You should check adjustment just about every time you use the machine, and the stops just get in the way. If you can figure out how to get them off and throw them away, you won't be sorry.

The best square I've found for setting machinery is a plastic 45/90 draftsman's square. (A steel one would be better if you can find one.)* It's the best thing there is for inside measurements like jointer fences, miter gauges, and radial saws, especially on the 45-degree side.

Six: Sharpening without Stones

You do it just like Grandpa used to do it: fish around in the drawer, find the stone—maybe it was Grandpa's stone—a little worn down in the middle, greasy . . . maybe you even have Grandpa's old oilcan,

*Currently available from Woodcraft (see sources).

also greasy, caked with black sawdust . . . give her a squirt, right in the middle, and lay on the steel.

You're not real sure how long you're supposed to rub the thing . . . there's a whole fishing and hunting season on the old Barlow, naturally a little rust and two nicks . . . round and around she goes, big circles, little circles, figure eights . . . in your mind the blade flashes, splitting hairs, lopping branches . . . women swoon, strong men weep . . . how's it look? A little scratched. Do the other side for a while; hum a few bars of something appropriate for the ancient ritual of sharpening. Scrape your thumb across the edge. Why do people do that, anyway? Who knows. Grandpa did it. They say if it'll scratch your fingernail . . . hmm. . . . maybe it needs grinding.

Don't forget to stick it in water . . . you have to stick it in water or you'll fry it. Right after you get this nick out. Oops, too late. Blue blade.

Gee, if you just had some of the right stuff, the stuff in the catalogs, the Arkansas stones, and slips for the gouges, and Washita (whatever that is), and India stones, and strops and steels—if you had some of that stuff and knew how to use it you could get an edge for sure, every time.

But you probably don't, so you probably won't.

Do you have some kind of a belt sander? And a buffing motor? Or something you can use for a buffing motor, like a drill press or a radial saw, or even a table saw? If you do, you can sharpen everything in the place with them, from the tiniest carving gouge to the double-

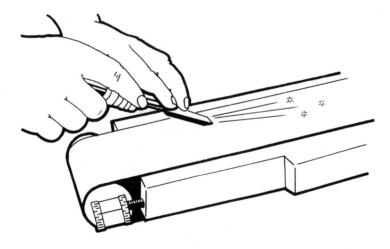

Hold the blade at approximately the original angle on a medium (80 grit) sanding belt with the belt moving AWAY FROM THE EDGE. Use light pressure and you can grind all day without burning.

bladed axe, so's they'll shave the hair off your arm (how else you gonna make women swoon?) every time; no stones, no slips, no oil, no burning, no fooling. You'll know exactly when the blade is sharp, and you'll do the job in half the time it takes with stones.

What you need to buy is a hard felt buffing wheel from a lapidary or jeweler's supplier or your local rock shop, and some emery buffing compound. If you have a 3450 rpm motor, a 3″ wheel will do. (You can use a conventional "rag wheel," but it cuts more slowly.)

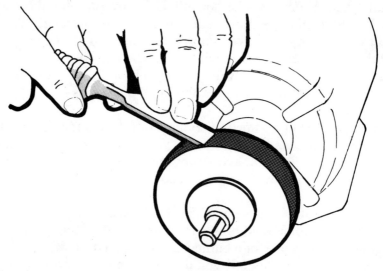

Buff the bevel to weaken the burr. Make sure the wheel turns away from you and *away from the edge*.

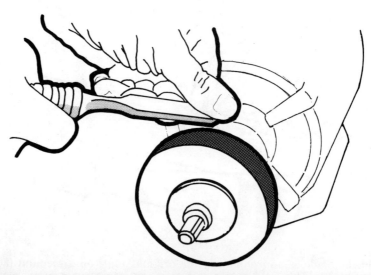

Flip the burr down with your thumb and back up again with the buffer. Keep buffing and flipping until the burr disappears. It only takes a few times. On knives, use the buffer on both sides.

Grind the bevel until the nicks are gone and a uniform burr appears all along the upper edge. On double beveled blades (knives) give both sides a lick.

As you move the burr back and forth, it will get smaller, and after four or five times it will disappear. "Strop" it a little on your thumb or a piece of leather to remove the buffing compound, and you're done. When the burr is gone, the blade is sharp.

Woodcraft Supply (see sources) sells a nice belt-sharpening system, although it uses a rag wheel which doesn't cut as fast as hard felt. The buffing compound they sell comes in a two-and-a-half-pound bar, and it works great. Two-and-a-half pounds will last so long you'll probably pass your first bar down to your grandchildren (instead of a greasy oilcan).

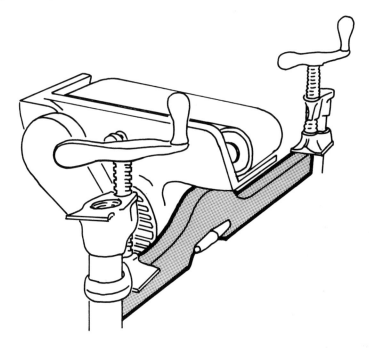

If you don't have a stationary belt sander, clamp your hand-held sander upside down with a wooden fixture. Be careful not to block air circulation.

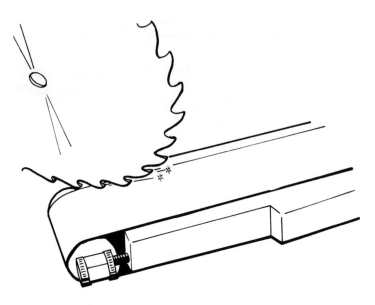

Do a quick-and-dirty job on circular saw blades. You can only grind the back of the tooth, and don't bother with the buffing wheel. Works fine, once you get the hang of it, but get the blade to a pro when you can.

"Slick-Ems"

For sharpening away from your power tools, make three wooden "slick-ems." Tack on a few layers of carbide paper—180 grit on one, 600 grit on another. Wrap the third in leather and rub in a little valve grinding compound if you can scrounge some (don't buy it; it's not that important).

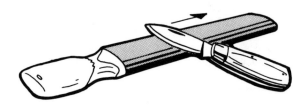

Draw blade down coarsest paper to form burr. Stroke each side a couple of times, moving burr back and forth. Repeat with fine paper until burr is gone. Strop lightly with leather.

POCKET "SLICK-EM"

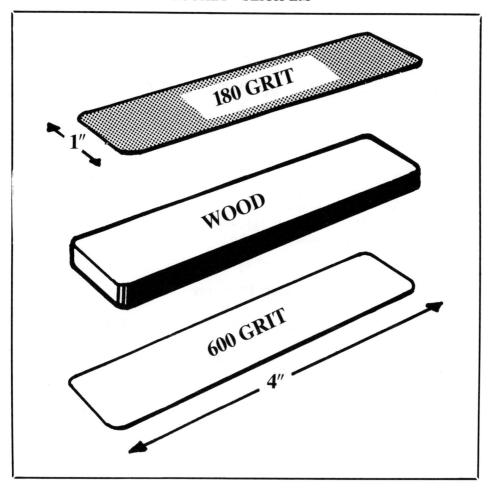

3
THE MOVEMENT

First order of business: go to chapter 10 (sources) and get the addresses of all the movement suppliers. You'll want to order at least the first two or three catalogs on the list.

One: Choosing

Hall clock movements are grouped into two broad classifications, popularly called grandmother and grandfather. Both do exactly the same job, but one is smaller than the other. The movement is smaller and the weights and pendulum are smaller. Movement size has little effect on dial size; you can use just about any movement with any dial, but the big movements are a little more rugged and less fussy. They're also more expensive.

The decision you have to make, once you've chosen the size, is basically a matter of chimes.

Rod Chimes

The most popular movements include a set of slender rods which are bolted to the case and struck by tiny, nylon-tipped hammers which are part of the chime trains of the movement.

The rods should be tuned when you get them, and permanently mounted in a casting, from which they should not be removed. Rod chimes come in three configurations: bim-bam, the least expensive, which plays no melody but counts out the hour like a cuckoo clock, with two-note "bim-bams" on the hour and a single "bim-bam" on the half-hour; Westminster, the most popular, which plays part of the Westminster chime melody (Big Ben) every fifteen minutes and counts out the hour; and multiple chime, which selects any one of two or three different melodies at the touch of a lever.

Tube Chimes

A tube chime movement is the same, basically, as the corresponding rod chime movement, except for the chime-striking mechanism and the price (tube movements cost nearly twice what rod movements cost). Tube chimes are long, inch-diameter brass tubes. They come either polished or chrome-plated. There are generally five tubes in a Westminster movement and nine in a triple chime. Tube movements seem to be currently available in only "grandfather" size.

The Difference

Rod chimes are bolted directly to the case. When they are struck, they cause the entire case to resonate, and the sound you hear is the sound of the case. It is the deep, mellow hall clock sound. Tube chimes are bells. They are suspended by string from their mounting, and do not couple their sound directly to the case. Tube chimes sound as good in free air as they do in a case—sometimes better. Since the sound comes directly from the bells, a tube clock must have an opening (usually in the sides or back of the hood) for the sound to come out.

A rod chime is like a stringed instrument; the sound is deep and mellow. Tube chimes are brass instruments; they are bright and clear. You takes your choice.

Other Movements

There are also "bell chime" movements available which strike the hour on a single brass bell mounted on the movement. These movements are sold as replacements for some antique clocks which originally had bell movements, but they have a unique sound and there's no reason not to use one if you like to be different.

And, of course, nowhere is it written that you have to use a chiming movement at all. However, if you don't, you won't have as many brass weights hanging down. Almost universally, the weight on the right (as you face the movement) drives the chime melody, the weight on the left drives the hour strike, and the one in the middle runs the movement. Bim-bam movements have only two weights (no melody) and non-chiming movements have only one.

Spring-wound movements work the same way: the winding hole on the right side of the dial is for the melody, the one on the left for the hour strike, and the one in the middle to make it tick.

Silencing

Make sure the movement you buy has some device for silencing the chimes. You may be used to the clock striking midnight, but your house guests won't be.

Two: Mounting the Movement

The movement, dial, and chimes will mount on a one-piece wooden carriage which will slide into the case from the front and sit on a bracket.

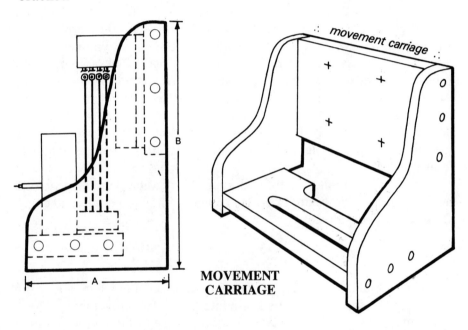

MOVEMENT CARRIAGE

Attach the movement to the carriage with the T-nuts supplied with the movement and hold the chime in place with a couple of small C-clamps. It's better not to bolt the chime in place just yet. It would be desirable to mount two shelf brackets on the wall in your "clean room" (any room in the house where there isn't too much dust) to set up the movement on so it can be running while you are building.

Under NO circumstances should the movement remain in the shop. Not even in the box.

Before you get into designing the case, you'll need to take some measurements on your movement:

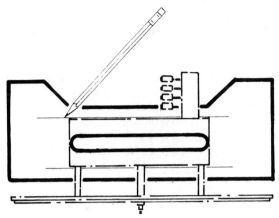

Make a seatboard that your movement will sit on, narrow enough to clear the chimes in back and the dial in front. Make it about as wide as the dial and cut a slot for the chains.

Mark the seatboard at front and back plates of movement.

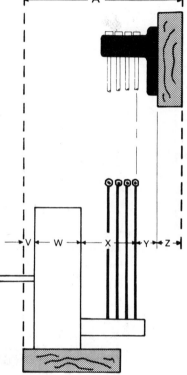

Total up all these measurements (v, w, x, y, z) and you'll have dimension A, the depth of the carriage side.

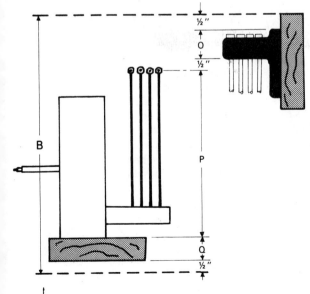

The total of these measurements (½+o+½+p+q+½) equals dimension B, the *height* of the carriage side.

It's a lot easier and more accurate to measure all these things individually and add than to try to hold everything in place and take one Grand Measurement.

33

MOUNTING THE DIAL

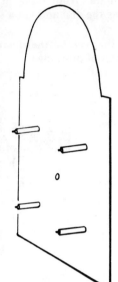

The easiest method for mounting the dial is on "feet," riveted to the dial by the movement supplier. The feet clamp into holes in the movement, so movement and dial become a single unit. The feet are attached to the dial to fit a *specific* movement, so order both from the same place if possible.

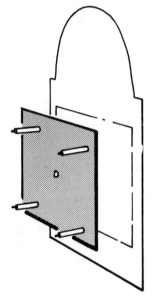

If necessary, you can buy feet, and, with careful measurement, attach them to your own dial, or a sheet metal backplate to epoxy to the dial.

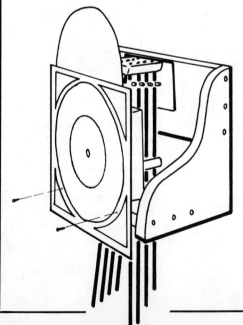

The dial can also be mounted by attaching it directly to the seatboard with small, unobtrusive brass screws, but ony if it's a very lightweight dial.

Screwing the dial to a plywood dialboard which is attached to the case is an often-used solution, but it can get unnecessarily complicated.

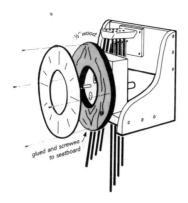

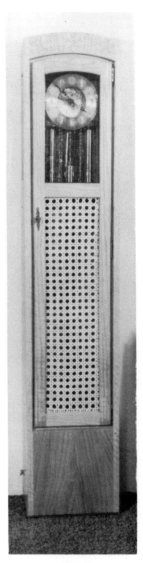

You may want to experiment with special dials. Here only the number ring is used; the movement is visible through the center.

"Koben"
—G. Williams

Another round dial application uses the entire square dial, but covers all but the number ring and the engraved center with a round-hole door.

"Narragansett"
—G. Williams

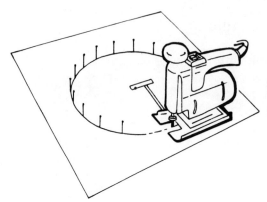

CUTTING A ROUND HOLE

Use a circle guide on a saber saw and insert snug finishing nails in the saw kerf, because when you finish the circle your center will break loose and cause a little jagged place in the cut. The nail spacers hold the center in place during that critical moment.

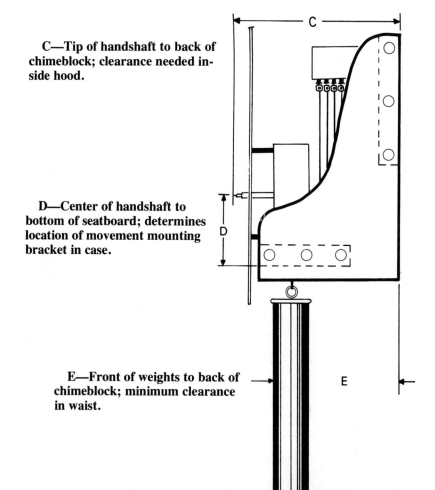

C—Tip of handshaft to back of chimeblock; clearance needed inside hood.

D—Center of handshaft to bottom of seatboard; determines location of movement mounting bracket in case.

E—Front of weights to back of chimeblock; minimum clearance in waist.

Three: Adjusting

Your movement should come with instructions for the few adjustments necessary to keep it running, so I'll just mention them briefly.

Crutch

If a pendulum movement does not tick evenly, it will stop running. It should tick evenly when the movement is level, but it's not always possible to have the movement level (the clock straight up and down) because in some houses the walls are crooked and the clock has to be set up crooked to align with the paneling, or a nearby doorway, or it will look funny. To compensate for all this crookedness, there is a crutch adjustment.

The crutch is a strip of sheet brass or steel, three to six inches long, on which the pendulum is hung. It clicks back and forth with no resistance, and each time it clicks the escapement moves ahead one notch. But click it to one side and then press a little harder and you'll feel it give; it's on a slip joint. Move it back to the other side and press, and it'll slip in that direction. That's the adjustment, and you'll probably have to make it every time you move your clock.

The first time you adjust the movement it will be on its seatboard, sitting on the shelf brackets you hung on the wall. Start it ticking. If the tick is not perfectly even and rhythmical, lift the right side of the seatboard a little. If the tick improves, press the crutch a tiny bit to the right and set the seatboard back down and see if it worked. If it gets better when you lift the left side of the seatboard, push the crutch a little to the left.

Pendulum Bob

This is an adjustment you can be making while the movement is hanging on the shelf brackets, because it takes a few days to get it right. If you've ever had a pendulum clock you probably know about moving the little nut under the bob up and down. Moving the nut (and bob) UP makes the pendulum shorter, and a short pendulum swings faster, so the clock speeds UP. Moving it DOWN and making the pendulum longer makes the clock slow DOWN.

Suspension Spring

This is not an adjustment, but it's something you should be aware of. You'll notice that the crutch hangs from a small, fragile rectangle of spring steel called a suspension spring. Be extremely careful not to kink this little rascal when handling the movement. If it ever breaks, you can replace it by pulling the taper pin or screw that holds it in place. In time, due to metal fatigue, the spring probably will break from normal use, so it isn't a bad idea to order an extra spring now and put it in a small envelope inside the case somewhere. They're cheap.

Chime Synchronizing

In modern movements we are blessed with very forgiving chime trains. You can spin the hands around, forward or backward, and get the chime all out of sync, and then leave it alone and in a couple of hours it will have corrected itself and chime according to what the hands indicate. Initially, however, the hands have to be set up according to what the MOVEMENT indicates.

Install the hands any old way and move them to the first quarter hour position and let the movement chime. Keep moving them, a quarter hour at a time, allowing the movement to finish each chime before moving on to the next quarter hour. When the movement chimes the entire on-the-hour melody and counts out the hour, take the hands off and set them up to read that time. The minute hand is on a square shaft; it will have to come off and be repositioned to the straight-up position. The hour hand is a snug fit on a round shaft. Set it to point at whatever hour the movement has just counted out. The movement and hands should now be in permanent agreement, and you should never have to remove them to synchronize again. When you change to daylight savings or something, just spin the hands around to the correct time, and in an hour or so the movement will catch up and start chiming correctly.

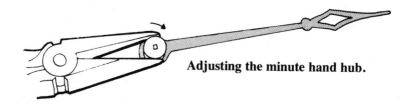

Adjusting the minute hand hub.

Minute Hand Correction

Very likely, when the clock strikes, the minute hand will not be pointing exactly at the twelve mark. It will be off a quarter inch or more in one direction or the other. Take the hand off and you'll notice it has a thick hub. This hub is a tight press fit on the hand, but it is movable. Grip the hub with pliers and move the hand in the desired direction and put it back on the clock. If you didn't get it exactly the first time, try again. It's best to stop the movement while you're doing this, or by the time you get the hand to say exactly twelve it'll already be a minute or two after!

The Handshaft Nut

The little nut which holds the hands on must only be tightened finger tight, so it may someday fall off and disappear into the vacuum cleaner. Don't despair; the movement suppliers sell them separately, as do some clock repair jewelers. Why not buy an extra one now and stick it in the envelope with your extra suspension spring? Nothing like being prepared.

Oiling

You shouldn't have to oil your movement for at least a couple of years. When you do, it is important that you oil *only* the shaft ends. Never oil the gears. You can see most of the shaft ends poking through the front and back plates of the movement. Dip a needle in clock oil (from the movement supplier) and touch the tiniest drop to the end of each shaft, front and back. It won't hurt anything if, before you oil, you obtain an aerosol can of tape recorder head cleaner, preferably with a long tube on the nozzle, and give the gears a blast to clean off any accumulated dust and oil. Head cleaner evaporates without residue, but be careful about getting it on the finish of your case.

Keep the little vial of oil in your spare parts envelope.

You'll know the movement needs oil when it starts running erratically for no apparent reason, or when it begins to slow down and you have to keep adjusting the pendulum bob. If blowing out the dust and oiling the shaft ends doesn't do the job, take it to an expert.

4
DESIGNING

If you are an average home woodworker, building a grandfather clock will very likely be a once-in-a-lifetime situation. Big clocks have a way of becoming the family's most treasured possession; they're traditionally built to be handed down, and they're expensive, even when you build them yourself.

All this makes building a clock something of a responsibility. The safest and most logical approach to design (and the one most often taken) is to buy a movement from one of the suppliers, add a dial that he says goes with it, and then order a set of plans from him for a clock which he has designed around this movement/dial combination. The advantages are obvious. First of all, all the figuring has been done for you. The plans are proven, and there is usually a photograph of the finished clock, so you can see exactly what you're supposed to end up with. Also, if there are any moldings or turnings in the design which you are not willing or equipped to make yourself, he can supply them. In fact, some of the suppliers of movements will supply parts for the clock in just about any stage of completion, right up to a completely assembled case, ready for putting on the finish.

This approach has its problems, however, the most important being, as far as I'm concerned, that what you end up with is somebody else's design. Personally, I attribute the joy I feel when I stand back and look at a new clock only partly to pride in the workmanship; much more important is the satisfaction of seeing the design I scribbled out weeks or months ago become a reality.

Another problem with buying somebody else's design is that in leafing through the catalogs your taste develops, you begin to know what you really want, and you may realize that what you want isn't really any of the designs, exactly; it's the crown of this one, the base of that one, and the overall shape of a third.

And there are other, more practical difficulties. Some of the plans are so complicated you have to be an engineer to read them, and sometimes, unfortunately, after you have painstakingly assembled your clock exactly to specifications and the stock swan's neck molding you've been waiting for arrives, you find out it doesn't fit, because the people you bought it from weren't as painstaking as you were.

So you'd like to design your own, but you don't think you know how, and you're worried that the weights will scrape against the door and the pendulum will bang back and forth against the sides of the waist and the dial won't fit. The answer is very simple. Design the clock from the inside out. Buy a movement and dial and build a box around it.

One: The Fixed Elements

A clock design develops around certain fixed elements. There are certain things which cannot be changed, around which the rest of the parts must be fitted. The fixed figure you have to start with is the money you can afford to spend on the innards. You can estimate pretty closely what you're going to have to spend on lumber for the size case you have in mind (figure high—remember what we said about the wood investment and the value of the finished clock; I usually estimate the amount of wood which will actually go into the case, and then buy twice that much), and you'll need a few shekels for glass and hardware and finish. If what you have left is between $150 and $200, you can buy a "grandmother" movement and dial. If you have $200–$500, you can buy a "grandfather" movement and dial. This decision is the first step in developing your design.

When you've decided on the movement and dial, buy them. You should have them both in your possession, hanging on brackets on the wall, before you cut your first piece of wood. It'll take two to four weeks for them to arrive, so spend your time doodling, getting the shape you want figured out, preferably on graph paper, so you can keep track of proportions. Don't worry about changing your mind several times while you're waiting for the movement to arrive.

The fixed shape that decides where the design begins is the dial. When you get the dial, trace it on a piece of paper and find the center of the arch by bisecting two chords (see below). The "break-arch" on

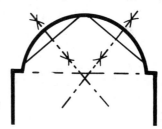

Find the center of the arch by bisecting two chords.

a clock dial is usually not an exact half circle; the center of the arch will often be a half-inch or so below the top of the square part of the dial.

Now reduce the outline of the dial onto your graph paper (I prefer ¼" graph paper—one square = 2 inches). From the shape of the dial comes the shape of the door, and from the door, the shape of the clock. (While I'll be describing a wide case here, most of the process applies to a narrow-waisted case as well.) Decide on the width of the door rails, usually about an inch and a half to two inches, and with your compass point still in the center of the arch draw the curve of the top of the door one-and-a-half to two inches above the top of the dial. Draw the rest of the door.

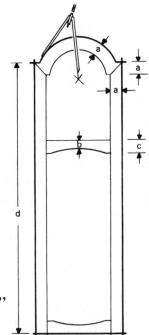

a. 1½–2 inches (all three "a" dimensions must be equal).
b, c, d. You decide.

Don't worry too much about allowing for rabbets or clearance or any of that now; you can make all those refinements when you make the patterns for cutting out the parts (see chapter 5).

At this point you have to decide on some kind of a crown. A crown generally consists of a board, which I call a facepiece, and a molding. Both the facepiece and the molding are mitered to wrap around the sides of the clock. The bottom of the facepiece will have an arch about ¾" above the top of the door. Use the same center. From here up is where your creativity gets a workout. The crown molding, the shape of the very top of the case, is what gives your clock the greatest part of its character. There are several more or less traditional ways to go, and of course you can invent your own. There is one thing to consider, however, before you attempt anything too radical: remember that a grandfather clock is a traditional piece, and in making departures from the traditional the designer encounters a thin line between the unusual and the bizarre.

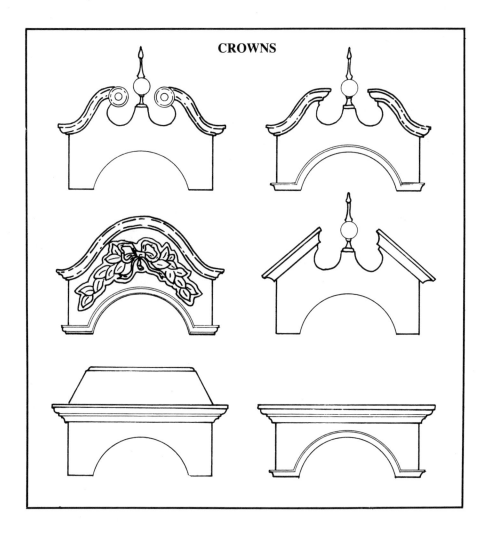

With the top and the door drawn, we are ready for the sides. The fixed figure which determines the width of the clock is the overall width the movement requires for the pendulum swing or the three weights, whichever is wider. The inside of the case should be around four inches wider than the width required by the pendulum swing. Overall width is seldom a problem with a wide case clock, so pick a dimension that you think will make for pleasing proportions; somewhere between eighteen and twenty-four inches for a six-and-a-half foot clock.

Height is essentially a matter of personal preference. However, one thing to consider is proportion between the movement and the case; the little twenty- or twenty-five-inch pendulum on a small movement would look like a necktie on a seven-and-a-half-foot clock, and a massive pendulum and huge weights knocking around in the base of a tiny case would look equally comical. As a rule of thumb, if you use a grandmother movement, stick to five-and-a-half to six-and-a-half feet. With a grandfather movement you can go six-and-a-half to seven-and-a-half feet or more. Just watch those eight-foot ceilings. A clock shouldn't just squeeze in under the ceiling—it'll appear crowded and out of proportion with the room. Give yourself at least a foot under the ceiling. (If you happen to have twelve-foot ceilings, have at it!)

Two: The Narrow-Waisted Case

While the traditional narrow-waisted case appears to be three boxes stacked on top of each other, it is very often only two boxes, so that the hood and the waist are the same size (*upper right*). Add a little molding and assorted other gingerbread and two boxes become three boxes!

It can be done other ways, but you're likely to get into complications with the movement and the moldings and a few other things if you build a separate top box which is very much larger than the waist.

You can build a top box with a ¾" offset as shown (*lower right*), but don't go any more than that; the fore-and-aft distance between the dial and the front of the weights isn't great enough. (Offset the front of the top box and the dial has to come forward, and with it the movement and weights. Get too far forward and the weights will not clear the waist door.)

A narrow-waisted clock is sketched out in about the same way as a wide case:
1. The dial width plus the width of the door rails plus the thickness of the sides is the width of the hood or hood/waist box.
2. Make sure the inside dimension of the waist is ample for the pendulum swing plus a few inches.
3. The base offset should be about the same as the hood molding offset (dimension "a" in the drawing lower right).

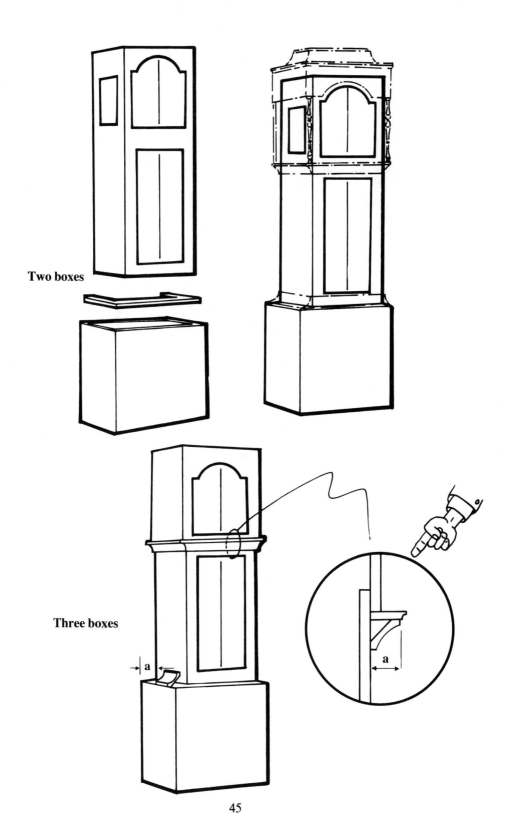

Two boxes

Three boxes

45

DESIGNING A GRANDFATHER CLOCK

1. Draw dial.

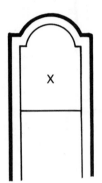

2. Decide width of door rails.

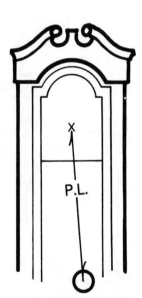

3. Decide overall width and sketch crown. Draw in pendulum length (P.L.).

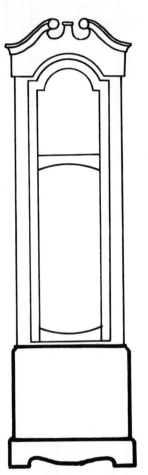

4. Choose overall height and sketch base.

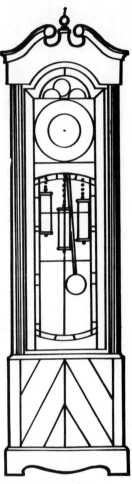

5. Fill in details.

4. The hood and base will be slightly taller than they are wide; the actual height depends on your own personal sense of proportion.

5. The height of the waist should be such that the pendulum swings somewhere in the lower half of the waist door. The pendulum length given in the catalog is usually measured from the center of the hand shaft to the center of a round bob. Cylindrical bobs are measured to the bottom.

Some suppliers will let you choose your pendulum length on some movements (Craft Products does), so pick a long one if you are planning to build a tall case.

Three: Notes on Design

• The early clockmakers rarely used glass in the waist door; presumably, to them the weights and pendulum were merely necessities, and not nearly as exciting as we think they are today. It is a fact, however, that there is seldom a visitor in my shop who can abide a solid door that "covers up all that pretty brass." You takes your choice.

• There are some nice non-brass dials available with painted designs and painted numerals on white. They are inexpensive and can work well in a homey design in pine or oak. Remember that the colonial American clockbuilders were often local cabinetmakers, and they worked extensively in pine and handpainted their dials. There is no disgrace in using pine (or any kind of real wood), as long as it's good quality pine and you don't attempt to make it look like something it's not. If there is a painter in your family, why not think about making your own dial?

• The wide case is a modern trend. The narrow waist is traditional, but so what? You takes your choice.

• Try to keep your design honest; a lot of stuck-on moldings and panels and doodads only clutter.

• Molded and pressed-wood carvings have no place anywhere, much less on a fine piece of furniture.

• Decide at the outset what sort of clock you're after—a kind of normal walnut-and-brass elegance, log-cabin pine, something clean and modern and striking, or whatever.

• Don't get gimmicky.

• Be aware of the limitations of your equipment. Large, curved moldings require larger shaper heads than are usually found in the

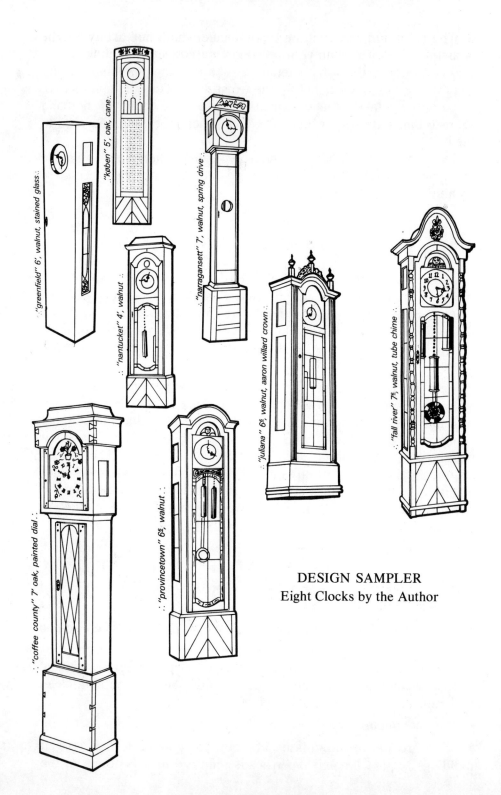

DESIGN SAMPLER
Eight Clocks by the Author

home shop. If you have to buy a molding you can't make, buy it right away, and have it in your possession before you start building.

• Try to find the originality which fits narrowly in between the straight traditional and the bizarre.

5
CONSTRUCTION

The Kuempel Chime Clock Works give two good pieces of advice in their plans (see sources), one of which is to build the door first. We designed the case from the dial out, because the dial is a fixed, invariable shape. It also makes sense to build the case from the dial out, starting with the door. Once the door is finished, we can build the rest of the case to fit it. Any corrections which may be necessary in door fit can be made in building the front of the clock, not in the door, because the door has to fit the dial. Sounds a little backward, but it's easy and it works. In practice, you won't have to worry about too much in the way of "corrections" if you cut everything carefully.

The other good advice from Kuempel is to make all the curved pieces (door top, crown, facepiece, etc.) of two angled pieces instead of one horizontal piece. The results are much more pleasing to the eye, and probably stronger. The drawings and photos which follow will proceed through the construction of a typical wide case grandmother clock. Everything has been made as uncomplicated and foolproof as possible in keeping with a high degree of quality in the finished project. The first step is making the patterns for the curved front pieces.

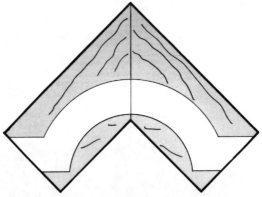

Use two pieces for the curved parts.

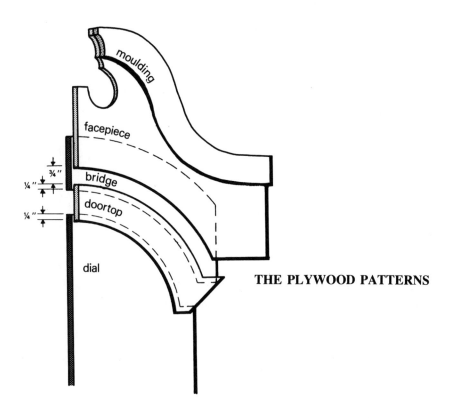

THE PLYWOOD PATTERNS

One: Making the Patterns

Lay your new dial down on a piece of ½" plywood or masonite and outline the top half of it. Find the center of the arch by bisecting two chords. It will probably fall slightly below the top of the square part of the dial.

DIAL

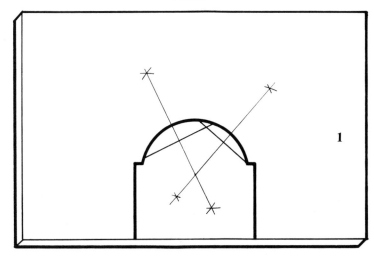

51

DOOR TOP

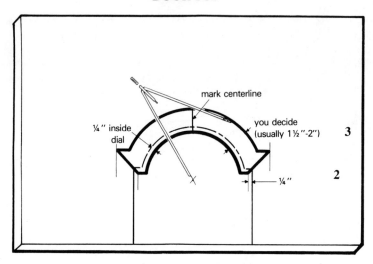

Draw and cut out the doortop pattern.

Draw the dial, doortop, and centerpoint again and lay out the bridge as shown.

BRIDGE

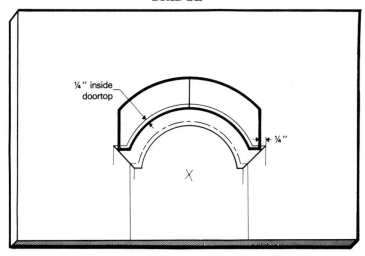

FACEPIECE

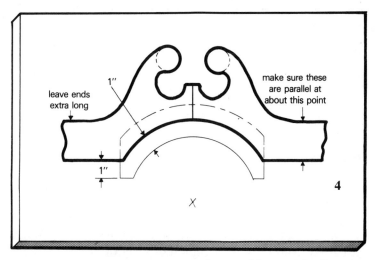

Trace the bridge pattern on a new sheet and lay out the facepiece.

Two: The Door

A long, narrow door is difficult to make absolutely flat and true. If you're not quite careful it will twist on you before, during, or after construction, and there's not a thing in the world you can do about the twist once it's there. But there's one thing you can do which will help prevent it, and that is to laminate the parts from two pieces of ½" or ⅝" stock.

WARPING

As wood dries out a couple of things happen. One thing is that the annual rings tend to try to straighten themselves out. When curved rings in a flat board try to straighten, the board become curved in the opposite direction, and you have a warped board on your hands.

Another problem is shrinkage. Dry unfinished wood takes on and gives off moisture as humidity conditions in the shop change. If you lay a wide board on the workbench and leave it for a couple of days, chances are it will cup up away from the bench. The reason is that moisture is able to escape from the top of the board, thus shrinking that side slightly, while the bottom loses less moisture because air cannot get to it, and it does not shrink. Often if you turn the board over so air can get to the other side it will flatten back out.

Laminate your door parts (all the parts, not just the long rails) from two pieces with the annual rings going in opposite directions so that warping stresses oppose each other.

like this . . .

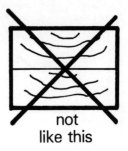
not like this

And then handle the parts with care, until the clock is finished; keep both sides of each piece exposed to the air as much as possible. This is a good rule to follow with completed parts of any project, and, as far as that goes, with wood in general.

Keep the pieces on thin "stickers," and support the whole board, not just the ends.

Cut the door rails out of the center of a nice symmetrical pattern . . .

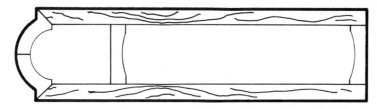

. . . then reverse them. Gives a nice mirror effect.

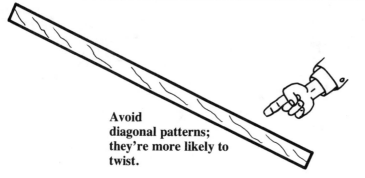

Avoid diagonal patterns; they're more likely to twist.

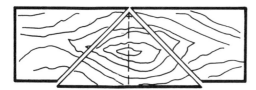

Make a "V" cut in the middle of a nice pattern for your door top . . . wastes a little wood, but it looks good.

CHOOSING GRAINS

Clamp the rails to your gluing jig (see chapter 2) for just an HOUR. White glue can usually be un-clamped in an hour. If you use a slower glue, cover the top of the rail with tape or wood before clamping so air can't get to it. Then remove and get them up on stickers.

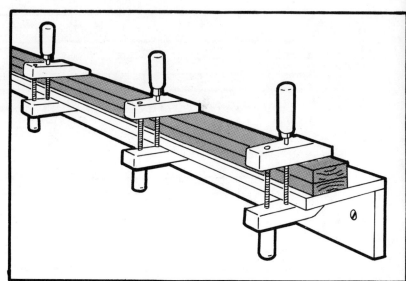

This spring clamp costs about $15. It's worth it. (See sources.)

Temporarily clamp two mitered pieces together and draw door top from the pattern.

Dowel them together. This Stanley doweling jig is available at larger hardware stores.

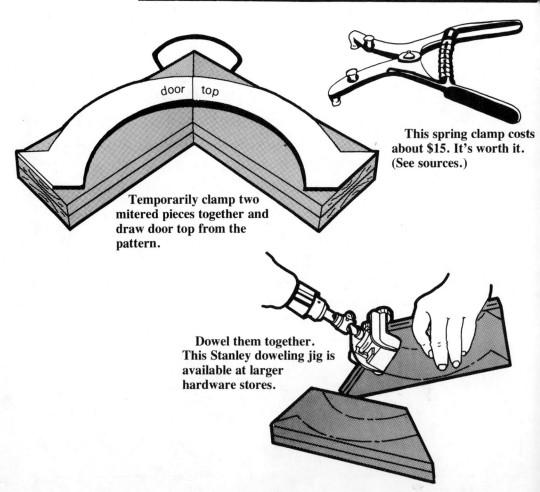

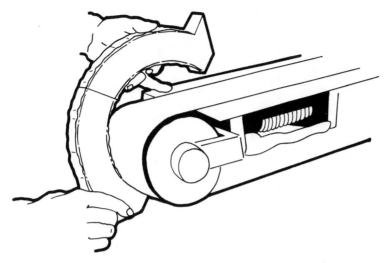

Bandsaw about 1/16" oversize and belt-sand right down to the line.

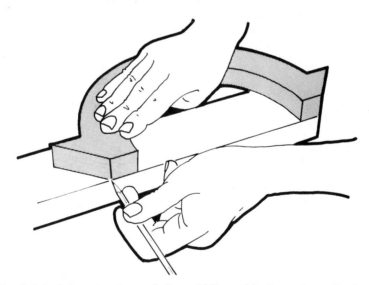

Use the finished door top to mark the middle and bottom pieces for length.

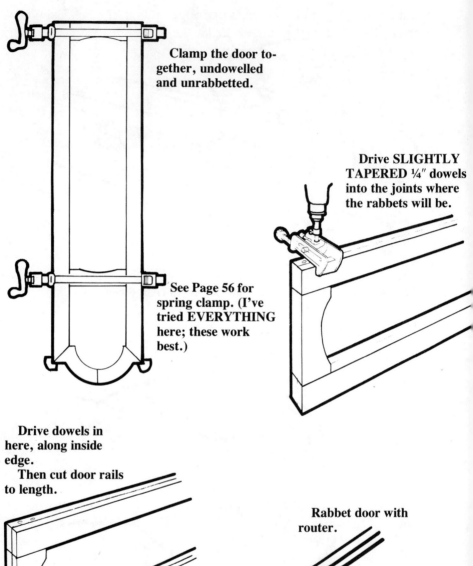

Clamp the door together, undowelled and unrabbetted.

Drive SLIGHTLY TAPERED ¼" dowels into the joints where the rabbets will be.

See Page 56 for spring clamp. (I've tried EVERYTHING here; these work best.)

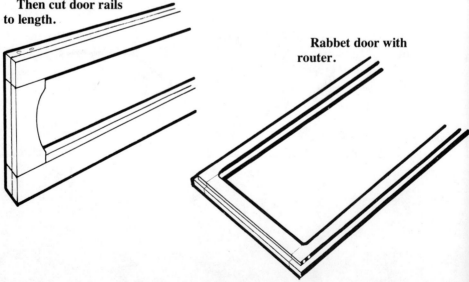

Drive dowels in here, along inside edge.
Then cut door rails to length.

Rabbet door with router.

Rabbet the assembled door, inner edge and outer edge, using a ⅜" rabbet cutting bit in the router. Make the rabbet ⅜" deep. Clamp the door to the bench or have somebody hold it.

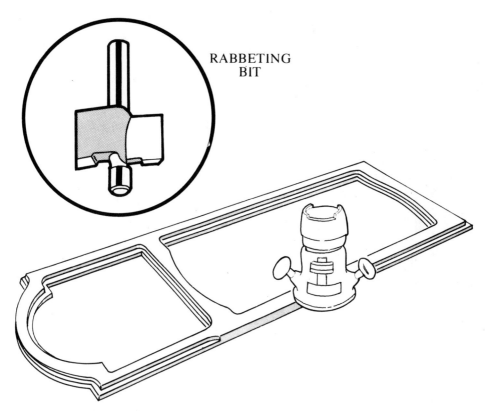

RABBETING BIT

Don't worry about the rounded inside corners; we'll round the corners of the leaded glass to fit. (If you don't use leaded glass, you're going to have to chisel 'em out, so be a sport, try leaded glass . . . chapter 9 shows how.)

Three: The Box

The basic frame of a wide-case clock is a simple four-sided box. Starting with the front it goes like this:

The Bridge

Lightly trace the pattern on the unglued pieces to determine location of dowels.

Dowel it together.

Back of door

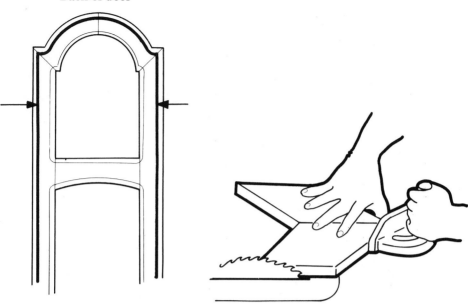

Measure from rabbet to rabbet on the door and add ¼" to determine width of bridge. Cut with the miter gauge set at 45 degrees.

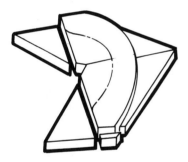

Bandsaw it out, but save the inside curve for the saber saw after the front is assembled. That little semi-circle of wood adds strength when you're clamping the front together.

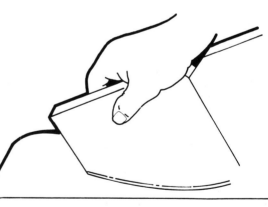

NOTE: A practiced hand on a belt sander is a lot more accurate than a bandsaw, *but* it's not as accurate as a circular saw. Never sand a joint to make it fit; re-cut it.

The Front

The front rails determine the final width of your case. They can be any width you like (within reason); somewhere around one-and-a-half to two inches is nice.

Dowel it together with the dowelling jig or glue it up and drive long dowels in from the rail edges after the glue is dry. They will be covered up by the sides.

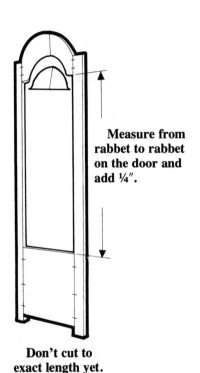

LONG DOWELS

If you have to drive exposed dowels in very far, use commercial turned dowels (with glue groove) or taper the dowels, or use a 1/64" oversize bit.

Measure from rabbet to rabbet on the door and add ¼".

Don't cut to exact length yet.

The Sides

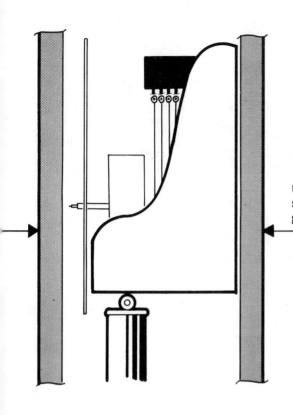

Make sure the side is at least this wide. Better add an inch or so just to make sure things don't get too crowded in there.

Don't cut either end to exact length.

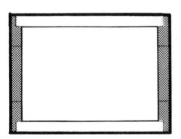

Sides will go on like this.

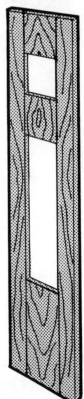

Try to cut the rails out of "mirror grains" as you did the door rails. Also, if you cut the three center pieces out of the same board they'll complement each other.

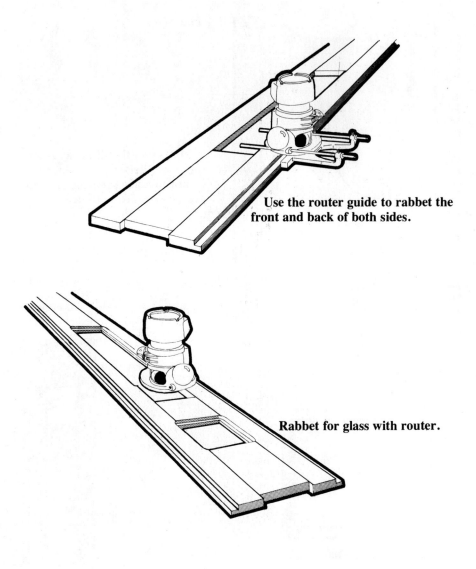

Use the router guide to rabbet the front and back of both sides.

Rabbet for glass with router.

NOTE:

If you're planning on getting craftsmanlike and mitering down the long joints between the front and sides, don't. It's a miserable clamping job, and besides, that kind of joint is structurally lousy. On a wide case you'll probably have some kind of a molding running down each side of the door; that molding is there to cover up a lap joint.

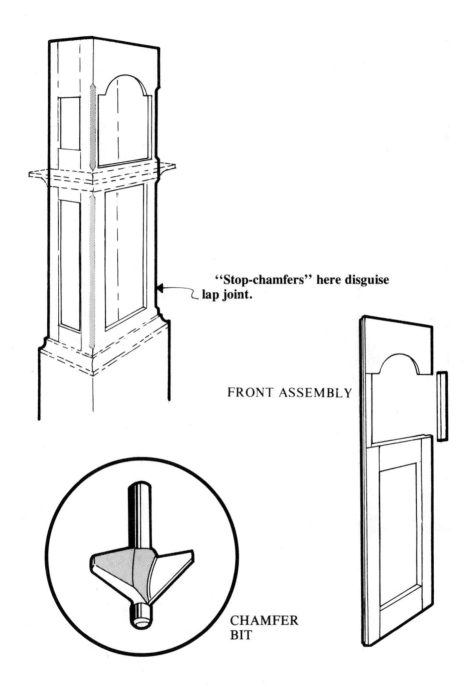

"Stop-chamfers" here disguise lap joint.

FRONT ASSEMBLY

CHAMFER BIT

If you're building a narrow-waisted clock, make the rabbets in the FRONT (so they show at the side) and then stop-chamfer them with the router. They'll look fine. A good lap joint looks a lot better than a puttied-up miter.

The Back

Cut up a bunch of your left-over narrow pieces to some equal width and glue up a panel slightly larger than you need. Get all the best surfaces on the front side of the panel if possible.

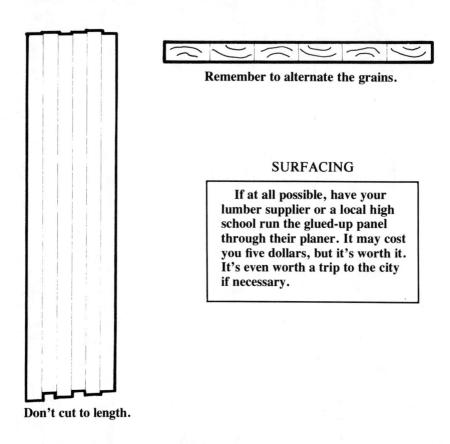

Remember to alternate the grains.

SURFACING

If at all possible, have your lumber supplier or a local high school run the glued-up panel through their planer. It may cost you five dollars, but it's worth it. It's even worth a trip to the city if necessary.

Don't cut to length.

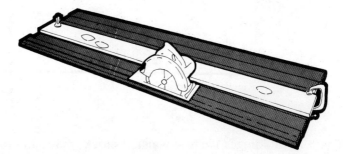

Using a plywood fence, rip approximately equal amounts from each side to make panel the exact width of the front.

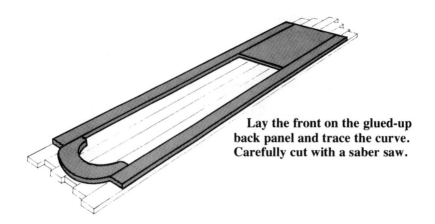

Lay the front on the glued-up back panel and trace the curve. Carefully cut with a saber saw.

PROBLEM:

How do you put a straight edge on those long boards?
Hardwood boards, when purchased in the rough or surfaced two sides, very often look something like this:

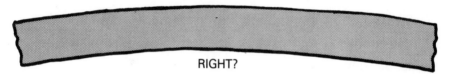

RIGHT?

Running an eight-foot board with a long curve across the jointer will usually just produce a prettier version of the same curve you started out with, because the jointer bed is short enough to follow the curve:

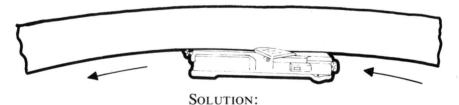

SOLUTION:

If you have the price of a small house or a large car, you can buy a thing called a "straight line saw," which will cut a straight edge on anything you can stuff into it, but if not, get a piece of string.

Strike a narrow chalkline on the board and eyeball a cut down the line with a skillsaw or saber saw. You should be able to stay at least within $1/16''$ of the line.

What this gives you is a somewhat shaky but straight edge, and *that* your jointer can handle.

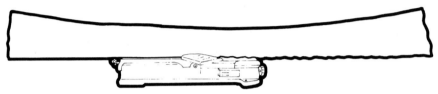

Once you have a straight edge, you can rip as usual.

NOTE:

If you don't have a jointer, you can rip with your hand-cut edge against the saw fence with pretty fair results.

NOTE AGAIN:

Wide boards are heavy and difficult to handle on the jointer, and for maximum resistance to warping, you shouldn't use them anyway. Make your hand cut through the middle of wide boards, with the chalkline or a plywood fence.

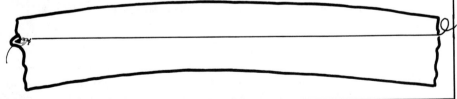

NOTE YET AGAIN:

Make yourself an end-holder-upper for ripping, if you don't have one, by simply clamping a scrap to a sawhorse.

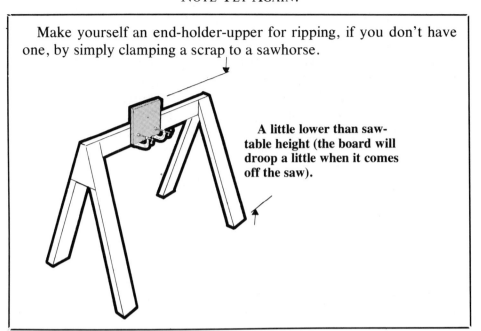

A little lower than saw-table height (the board will droop a little when it comes off the saw).

Four: Putting It Together

NOW cut everything to the same length.

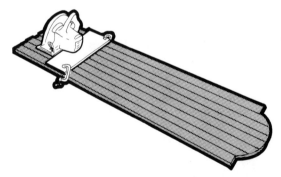

Nothing works better than the good old skillsaw and fence. Dry fit the box together to make sure everything is exactly the same length.
NOW STOP.
Before the glue starts flowing, sand the inside of all four pieces, all the way down to 220 grit. In the finishing chapter I recommend you belt sand with 60 grit where necessary, and then proceed by orbital or by hand from 60 grit through 80, 100, 120, 150, 220, and 600 wet/dry.

It sounds like a lot of sanding, but believe me, you do not save time by skipping grades. You save time by using only *brand-new* sandpaper.

Do the back with a new piece of each grade and discard it. Use a second new piece for the front and sides. Sandpaper is expensive, but it's not that expensive. Lavish use of sandpaper on the whole project will only cost you a few dollars, and you'll save that in backbreak. (Besides, you don't really throw away all those practically-new sheets; put them in a box and save them for the Cub Scouts, or use them when you need to hand sand a doodad.)

Now put it together. If you have enough clamps, glue the front and back at the same time. If not, glue up the back and clamp the front on dry with just a couple of clamps to help hold things square.

SQUARING UP

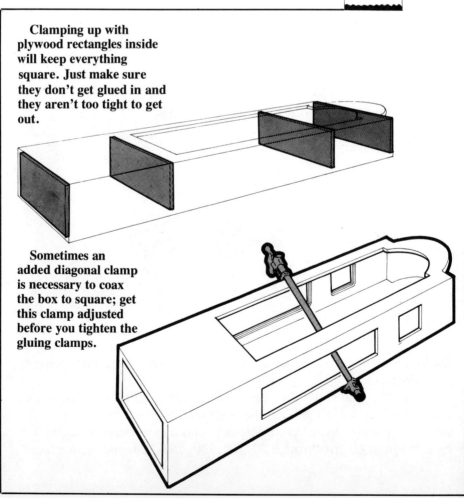

Clamping up with plywood rectangles inside will keep everything square. Just make sure they don't get glued in and they aren't too tight to get out.

Sometimes an added diagonal clamp is necessary to coax the box to square; get this clamp adjusted before you tighten the gluing clamps.

NOTE:

> Wipe all excess glue off with a WET rag as soon as possible. Keep a wet rag on hand whenever you're gluing; it comes off a LOT easier when it's wet than when it's dry!

NOW SAND AGAIN.

Sand the outside of the box all the way down to 600 now while you have flat, unencumbered surfaces. You'll have to touch it up again later, but it's important to get the worst over with now. Be careful to keep the surfaces flat and especially careful NOT TO ROUND THE CORNERS.

Five: The Roof

Now comes the only piece of plywood in the clock. If you're going to have a curved roof, you need a ⅛" panel, and plywood does the job better than solid wood. (⅛" solid stock is too fragile; it won't hold up with age.) Obtain a piece of ⅛×12×24 hardwood plywood from Craftsman Wood Service (see sources). It comes in walnut, mahogany, and basswood.

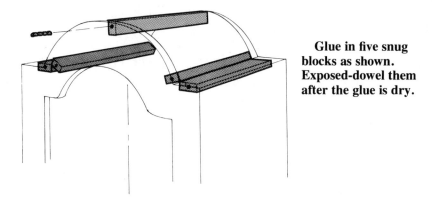

Glue in five snug blocks as shown. Exposed-dowel them after the glue is dry.

Put the plywood in the bathtub to soak an hour or so. Don't weight it down with anything metal; it'll leave a black spot.

Obtain some attractive nails. I use square cut "boat nails" from Tremont Nail Company (see sources). They are square, and they have a raised, antique head.

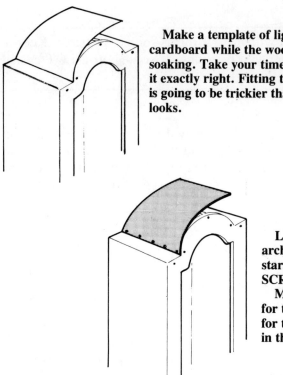

Make a template of light cardboard while the wood is soaking. Take your time and get it exactly right. Fitting this roof is going to be trickier than it looks.

Lay a bead of glue on the arches and attach the plywood starting at one end with small SCREWS.
Make sure the holes you drill for the screws won't be too big for the nails you're going to put in those holes later.

Put the lid on with screws and you can remove them when the glue is dry so you can sand and finish. When the case is all finished, you can tap the antique nails in the screw holes without the considerable banging they require if not pre-drilled.

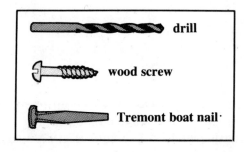

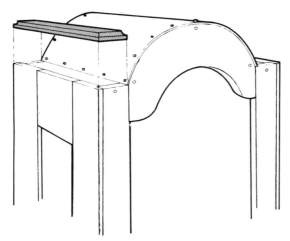

Shoulder Caps

Make some shoulder caps from about $5/16''$ stock with some kind of a shaped edge. Before you install them, sand them and the roof completely.

Six: The Bottom Box

Be creative down here, but don't get gimmicky. Look at all the cases in the catalogs and in the books and stores and get some idea of what you want. There are all kinds of things you can do with moldings and such, but this is the place where the temptation is greatest to buy something and stick it on.

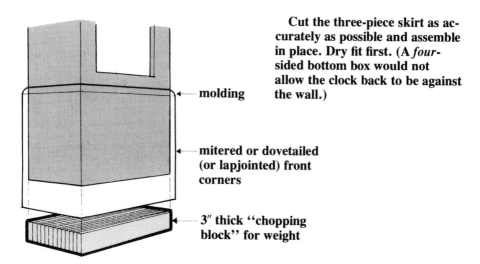

Cut the three-piece skirt as accurately as possible and assemble in place. Dry fit first. (A *four-*sided bottom box would not allow the clock back to be against the wall.)

← molding

← mitered or dovetailed (or lapjointed) front corners

← 3" thick "chopping block" for weight

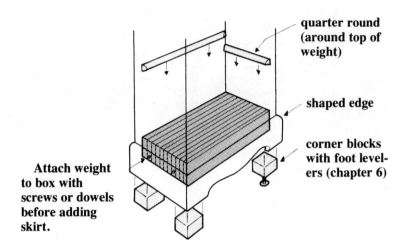

Don't insult your own workmanship by pasting on one of those molded carvings. If you wanted genuine-simulated-wood-grain-effect you wouldn't have come this far.

I would suggest *real* wood carving (chapter 9), marquetry, if that's your style (Constantine is a good source for all that stuff—see sources), raised panels, or diagonal grains (chapter 6).

For a small skirt, put the weight up inside the box and put quarter round around the top of it. The weight should be just a snug fit.

6

THE GINGERBREAD

At this point we have a box, or two or three boxes, sanded completely inside and out. Again, let us hope you didn't round the corners.

In adding the ornamentation, sequence is not too important, so what follows is a collection of individual procedures from which you can pick and choose.

One: Foot Levelers

Buy them from one of the mail-order houses (see sources) and install them by just drilling a $^5/_{16}''$ hole in each corner of the chopping block or corner blocks. You'll be glad you have them.

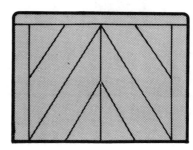

Two: Diagonal Grains

Cut a paper or plywood pattern half the size of the panel you want to make. Trace it at opposite angles on two glued-up pieces.

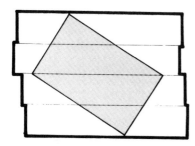
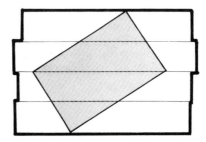

Glue up as many pieces as you like (I use three or four in each half) and make them any width you like. Just make sure they're the *same* width. Choose nice grains.

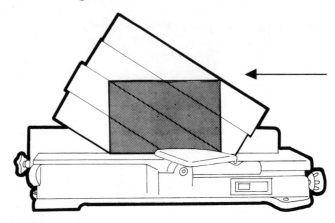

Carefully eyeball a cut 1/16" outside the line on the bandsaw, and then joint it down to the line. Do the same to the other half and rip them to exact width.

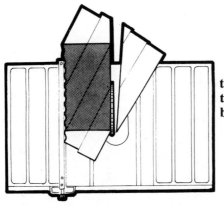

If you don't have a jointer, rip with the bandsawed edge against the fence, then reverse and rip away the bandsawed edge.

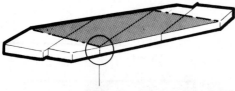

Find the point and make the crosscut there on both halves.
The points will then line up when you dowel the two halves together.

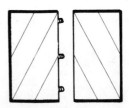

Three: Large Cove Moldings

Either of these tried-and-true methods will make all the cove moldings you care to hand sand, and hand sand you will, because you'll want to use a blade with a fair amount of set. A cabinet scraper with a curved end would be a good investment. TAKE SMALL BITES when making this molding; increase the depth of the cut about ⅛" per pass. You can buy all kinds of moldings from the movement/kit suppliers if you prefer.

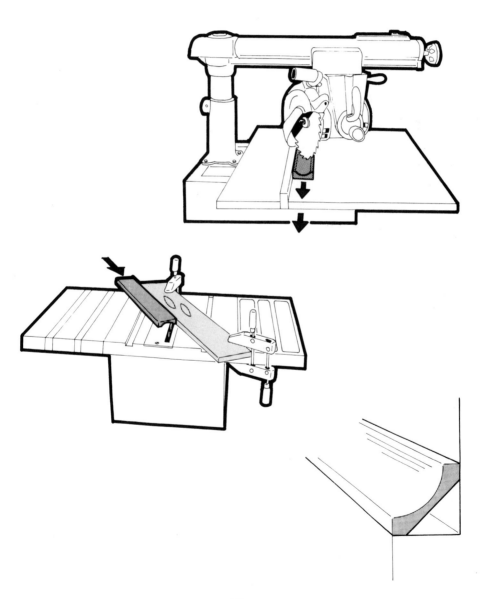

Four: Mitered Moldings

When gluing on a three-piece mitered molding, there are some tricks to help keep you out of trouble.

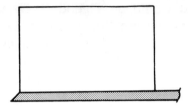

Miter one end of the front piece and mark for miter on the other end. Cut it and glue it on.

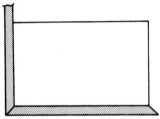

Miter one end of an extra-long side piece and dry fit. If it fits perfectly, pat yourself on the back, cut it to the proper length, and glue it on.

If it doesn't fit (nobody's perfect), fudge as follows:

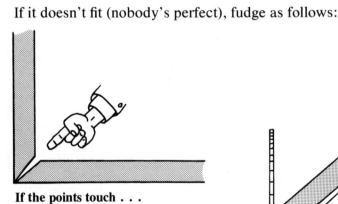

If the points touch . . .

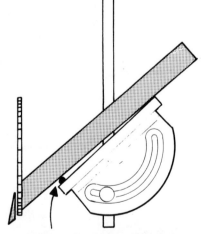

. . . stick a sliver here. Moving the sliver toward the center of the miter gauge will cut off more of the point.

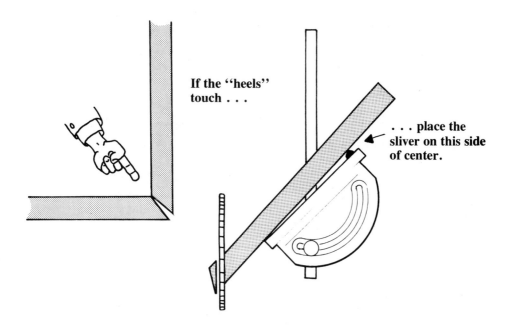

If the "heels" touch . . .

. . . place the sliver on this side of center.

These drawings are all WILDLY exaggerated, of course. You can make much more precise adjustments this way than by trying to adjust the miter gauge, but it's still essentially a trial-and-error business. That's why you cut the piece extra long to start with. When you get it right, then cut it to length and glue it on.

Five: The Front Moldings

The front and sides of the box are joined with lap joints. We made the joints show on the front because that's a natural place for moldings to cover them up.

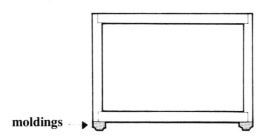

moldings

Buy stock moldings or make them yourself, but keep a couple of things in mind:

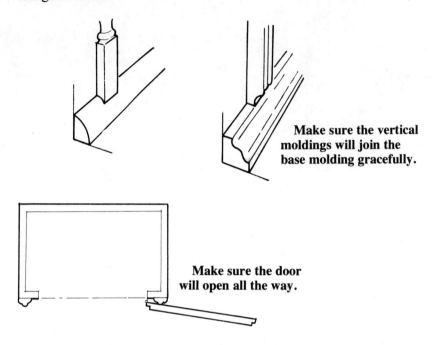

Make sure the vertical moldings will join the base molding gracefully.

Make sure the door will open all the way.

Six: Split Turnings

Split turnings make nice vertical moldings for both narrow-waist and wide-case clocks. Turning will be easiest if you cut the stock into short lengths—twelve to eighteen inches—and easier yet if you have a "steady rest" for your lathe.

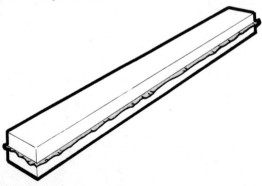

Glue the strips together with paper between, as you do for faceplate turning. Fairly thick paper will make separating easier.

If you're using several short pieces to make up a long molding, turn the ends to be joined so that they are slightly *concave*. It'll make the joints less detectable.

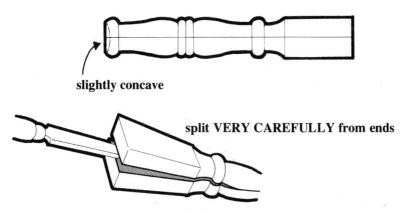

slightly concave

split VERY CAREFULLY from ends

Glue the sections on to form a continuous long molding. It's not necessary to make all the sections identical; it's more interesting if you don't. Just number both halves of each turning and make sure they are in the opposing positions on the case, so that the halves of no. 1 turning become the topmost sections, the halves of no. 2 go right under them, etc.

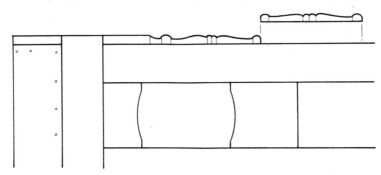

Seven: Swan's Neck Molding

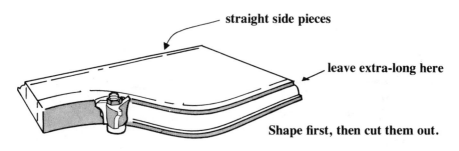

straight side pieces

leave extra-long here

Shape first, then cut them out.

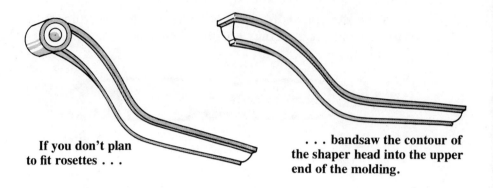

If you don't plan to fit rosettes . . .

. . . bandsaw the contour of the shaper head into the upper end of the molding.

Eight: Raised Panels

I can't say enough for raised panels. Learn how to make good raised panels and you can make all kinds of things out of solid wood. This kind of construction is not only traditional and very attractive, but it allows for the release of all the usual stresses in the wood, resulting in an extremely stable and long lasting panel.

There are all sorts of ways to "raise" the panels; the object of doing it, of course, aside from its good looks, is to make a ½″ to ¾″ panel fit in a groove in a ¾″ rail. This allows for large panel assemblies of many individual components, each "doing its own thing" with little effect on overall panel stability. (Any solid piece of wood *will* do its own thing, no matter how tightly you glue it, screw it, nail it, or whatever.)

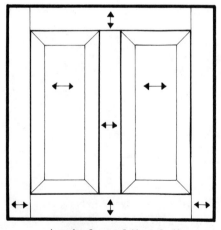

A raised panel "works" minutely in all these places. Overall dimensions affected only slightly.

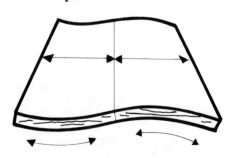

Panel of large boards works more in width; also warps.

RAISED PANELS ARE STABLE

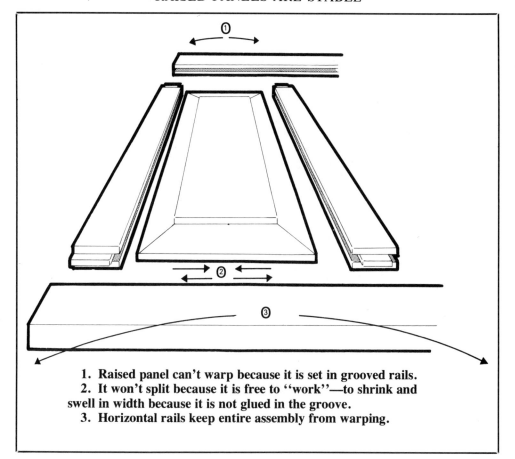

1. Raised panel can't warp because it is set in grooved rails.
2. It won't split because it is free to "work"—to shrink and swell in width because it is not glued in the groove.
3. Horizontal rails keep entire assembly from warping.

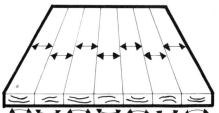

Panel of alternating grains tends to warp less because the stresses cancel each other, but shrinkage and swelling in width is additive.

We can get away with a glued-up panel for a clock back because the entire back is a single, vertical-grain piece. It won't split because there isn't really anything to keep it from working. The sides of your case may flex in and out microscopically, but you'll never know it.

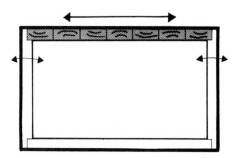

WHILE WE'RE ON THE SUBJECT...

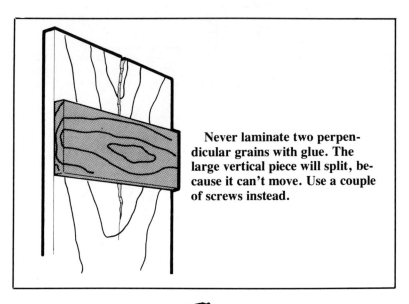

Never laminate two perpendicular grains with glue. The large vertical piece will split, because it can't move. Use a couple of screws instead.

Making the Rails

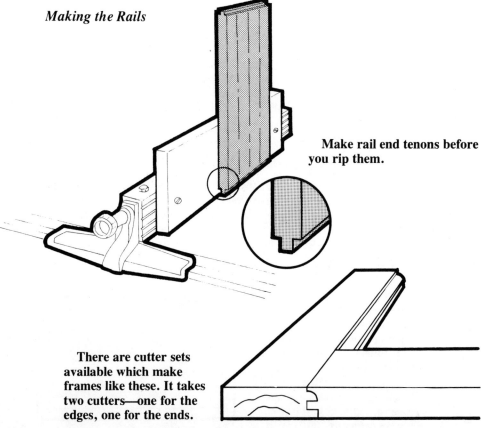

Make rail end tenons before you rip them.

There are cutter sets available which make frames like these. It takes two cutters—one for the edges, one for the ends.

If you use these cutters on the table saw, cut the rail ends very carefully. Take *tiny* bites—raise the cutter only a little for each pass. Do all the rails at once. Check Sears for a thing called a "Universal Jig;" it makes this job easier.

RADIAL ARM SETUP FOR CUTTING RAIL ENDS

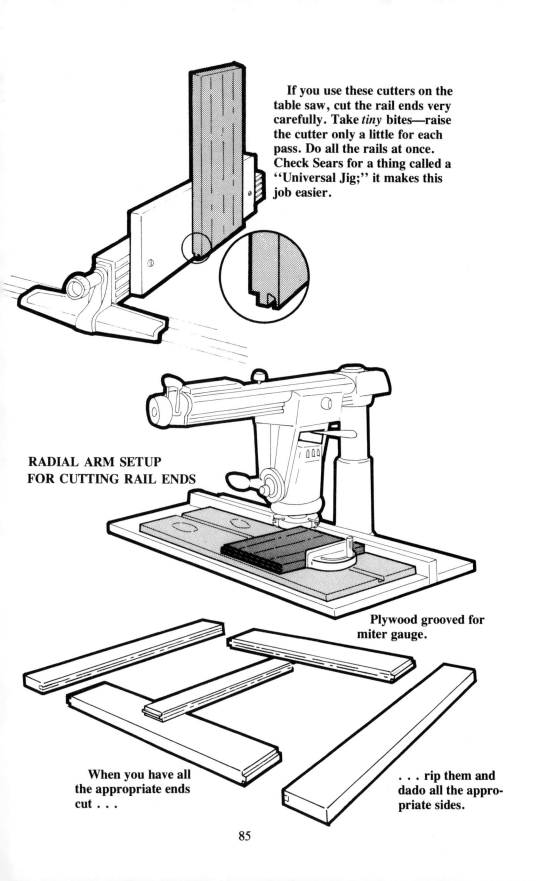

Plywood grooved for miter gauge.

When you have all the appropriate ends cut . . .

. . . rip them and dado all the appropriate sides.

NOTE:

> If you plan to cut the tenons with a dado head, set up the blades you will use for the edge groove first (two blades and a thin cardboard spacer for example), then cut a sample groove in a piece of scrap, so you will have something to check your tenon setup with.

Cutting the Tenons with a Dado Head

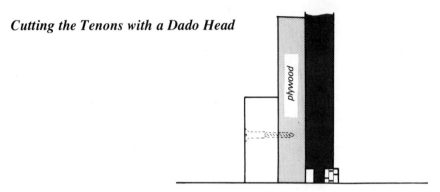

Experiment on scrap *the same thickness* as your rail stock until you come up with a perfect tenon.

Raising the Panels

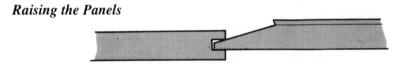

Because of the panel's taper, its thin edge will have to be thinner than the dado width. Experiment with scrap.

Table Saw

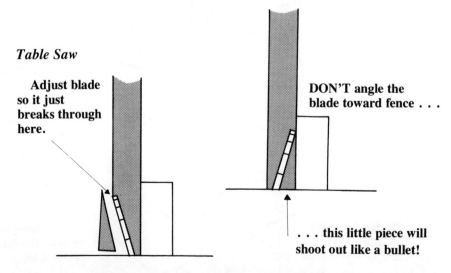

Adjust blade so it just breaks through here.

DON'T angle the blade toward fence . . .

. . . this little piece will shoot out like a bullet!

If the slot in your table insert is big enough so that the thin edge of the raised panel tends to fall in, make a blank one out of plywood, hold it down with a large scrap, then carefully crank the blade *all the way* up through it. Then lower the blade to desired height.

Radial Arm Setup

Use a small planer blade if you have one; saves a lot of sanding.

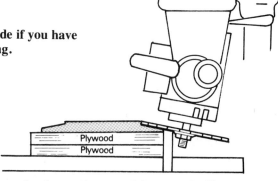

Once you have your raised panel assembly, use it like any other panel:

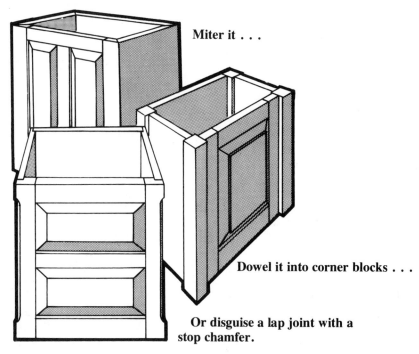

Miter it . . .

Dowel it into corner blocks . . .

Or disguise a lap joint with a stop chamfer.

Nine: Hinges

For simplicity, we made a cabinet-door-lip door, but you probably won't find a cabinet-lip-type hinge in the hardware store suitable for a fine hall clock. I like to use a semiconcealed brass hinge like the one shown below. Only the joint shows outside the case. They're a little hard to find (try Craftsman Wood Service or DeCovnick).

Mount the hinges to the door first. Then, with the case on its back, lay the door down in place and mark the case for hinge mortises. Make sure there is at least 1/8" clearance at the bottom of the door—leaded glass is heavy; a little settling is likely to occur.

Ten: Locks

I use a simple little no-mortise lock from H. DeCovnick.

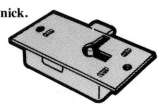

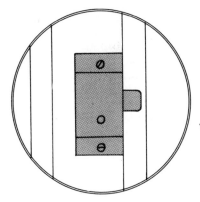

Just stick it on the back of the door rail with two wood screws.

The key shaft sticks out a little. Put the lock where you want it and press. The shaft will make a dent in the wood . . .

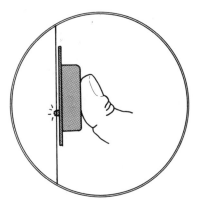

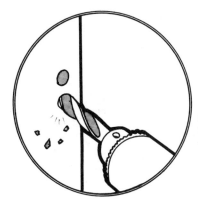

From the dent in the back of the door, drill a ¼" hole through the rail, and from the *front*, drill another ¼" hole right under the first one.

Now enlarge the top hole to 5/16", connect the two holes with a coping saw, and there's your keyhole.

(We could have drilled the top hole 5/16" to start with, but the larger drill is more likely to cause a little splintering as it comes out the front of the door.)

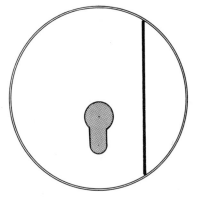

Attach the lock. Be sure to pre-bore for the tiny screws. You're committed to this location now; if you break off a screw by forcing it into too small a hole, you're cooked.

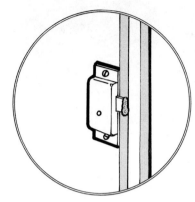

Turn the key in the lock to move the bolt out. Coat the end of the bolt with Vaseline, lipstick, soft wax, or some such. Click the bolt back in and close the door. Then try to lock the lock. The bolt will press against the wood frame of the case, leaving a nice mark right where you need to cut the bolt mortise.

Drill a few ⅛" holes and connect them by wiggling the drill up and down or with a small woodcarving chisel.

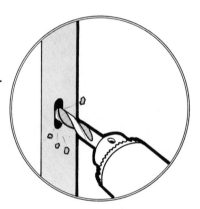

Eleven: Installing the Movement

The Bracket

Make a bracket as shown, an easy fit into the bottom of the movement carriage. Note that the grain in the back piece must run *up and down*.

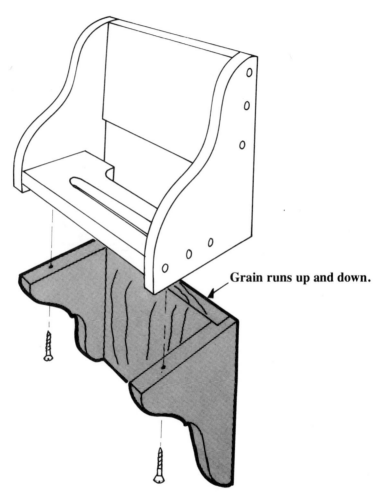

Grain runs up and down.

The bracket will be glued solidly to the back of the case, and, to avoid splitting the back, the grains should not be perpendicular. (See "While we're on the subject," page 84.

Measuring

The movement bracket must be located precisely so that the dial fits exactly in the opening of the dial door. It's easily done:

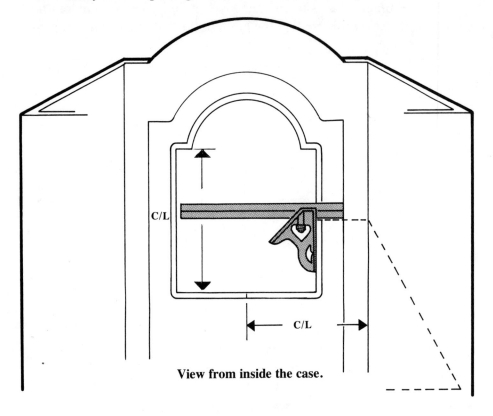

View from inside the case.

1. With the door hung on hinges in its permanent position, find the vertical center and transfer this line around to the back with a combination square.
2. Find the horizontal center and, by measurement, transfer this mark to the back.

This will give you crossed marks on the inside of the back exactly corresponding to the handshaft in the middle of the dial.

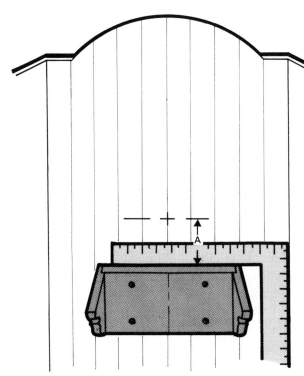

From this center mark, measure down dimension "A" (see below). With a square, strike a line across the back of the case here.

This new line marks the top of the bracket you just made.

Left-and-right placement of the bracket is determined by matching up the center lines.

Measure from the center of the handshaft to the bottom of the seatboard.

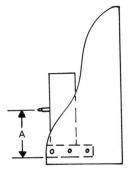

7

THE FINISH

In general, there are two kinds of finishes for wood: those which go ON the wood (varnish, shellac, urethanes, etc.) and those which go IN the wood (the oils). Surface finishes require a little less sanding, and they provide more protection than the oils, but when they get old and/or damaged they just about have to be stripped off entirely and a new finish applied. They can also be tricky to apply professionally, as I'm sure you know, what with dust and brush strokes and what-all.

Danish oil is pretty sensitive to moisture (it's not very good for table tops) but it eliminates most of the application problems, and when it gets damaged, it only requires a little sanding and a little more oil rubbed in with a rag.

If there is a drawback to Danish oil, it is that it adds no gloss to the wood; oiled pieces are usually satiny dull, maybe too dull for your liking. In fact, Danish oil doesn't add anything to the wood (after all, you wipe most of it off after you put it on); all it does is bring out the color and polish which were in the wood in the first place and protect it somewhat from the moisture and dirt of handling over the years. One of the major Danish oil manufacturers, Watco, claims that the oil also hardens the wood as it cures.

Watco Danish oil is the finish I use on clocks, but I don't get a dull, satin finish; I get a very satisfying highly polished finish.

The polish doesn't come from the oil—it's in the wood—and once you've seen really polished wood, darkened with clear oil, the gloss of a surface finish looks painted on forevermore.

One: Sanding

Since the oil doesn't really add anything in the way of polish, all the work is done to the wood itself. The "finish" is obtained with the sander, before the oil can is even opened.

Sanding is one of those jobs which can often be done better by machine than by hand, so do like it says in chapter 5: sand when it can be best done by machine—when you have large flat surfaces, before the molding goes on (and the doors, etc.)—and sand the insides of things before assembly.

Belt Sanding

If you have edge-glued two boards together, as you did with most of the clock case, your first concern is getting the glued-up panel flat. Even if you ran the glued-up back through a planer, you'll probably still want to start off with a belt sander. It's true that a belt sander can do irrevocable damage to your workpiece if you're not careful, but don't be afraid of it. Just keep it moving, don't let it tip, and don't overdo it. Remember that the object is to get the piece flat, and the best way to do that is to use coarse, 50 or 60 grit belts and SAND ACROSS THE GRAIN.

Forget what they told you in high school; you have to remove a lot of wood from the high board to bring it down level with the low board, and you want to lower the whole board, not just "feather it in."

Sand across the grain, concentrating on the high places, until you have it reasonably flat. Then sand the whole panel equally across the grain, then diagonally in two directions, and finally with the grain, sanding the whole panel evenly in each direction. Check with a straightedge and sand a little more where necessary. Get the thing flat. Don't worry about belt sander marks; we'll take care of those.

Orbital Sanding

If you read chapter 2, you read my ravings about the Rockwell commercial sander, but if you already have a sander you're probably not going to run out and drop a hundred dollars on another one just because I take mine to bed with me at night, so do what you can.

The most important thing to remember is to use new sandpaper. Sandpaper is too cheap and your time is too expensive for you to be fooling around with used sandpaper. On an orbital machine, fresh paper probably removes more wood in the first two to five minutes of its life than it will in the rest of its life, even if you use it until it disintegrates. And I'm not talking about flint paper. Flint paper is worthless. I'm talking about garnet (the orange stuff) or aluminum oxide (gray) cabinet paper. Use it lavishly; change paper on your sander every two to four square feet you have to sand. It'll not only make the job easier for you, it'll do it better, because sharp paper won't polish the high spots—it'll cut them off.

NOTE:

> Between grits, blow, brush, vacuum, sweep, and/or tack cloth the wood very clean; feel it with your hand and make sure you haven't left any bits of coarse grit to roll around like rocks under succeeding finer papers.

Start with 60 grit, and move the sander evenly over one whole side until the belt marks are all gone. Then leave the clock where it is and go on to 80, 100, 120, 150, and 220. Then turn the case to another side and take that side down all the way to 220. Saves a lot of wear and tear on your back and the clock if you don't move it around too much.

Remember you're going to be putting mitered moldings around the box, so be extremely careful not to round the corners.

Two: Polishing

When you have the case down to 220, go ahead and assemble it. It won't do any good to polish it now, because you'll probably be wiping glue with a damp rag and handling the case a lot, both of which will kill the polish.

Try to do the polishing on a dry morning, the day you plan to apply the oil. It wouldn't hurt to go back over everything with 220 again, hand sanding WITH the grain up against the moldings first, then with the machine. Then sand with 400 wet/dry carbide (black) paper. If you can't get it at your hardware store, try an auto parts house.

Work with a light source across the case from you; rig a trouble light or lamp if necessary. It helps to be able to see the reflections begin to appear in the wood as the polish comes up. Hand sand the moldings with wet/dry paper (you use it DRY, of course) or use 000 steel wool.

The final polish comes with 600 wet/dry paper. You don't want to change this paper too often, because in the first minute or so it is sanding, and it's not until it gets filled that it starts polishing. Use a new sheet on one entire surface of the case. Go over the whole surface evenly while the paper is still sanding, and then pull up a chair, settle into a good daydream, and polish.

Hold the sander in one hand and let it work with its own weight. Move it around in circles, as though you were polishing a car. You'll begin to see the light across from you and the sander reflected in the bare wood. Stay with it up to five or ten minutes per square foot. Use

000 or 0000 steel wool on the areas you can't get with the machine. Go with the grain when polishing by hand.

Clean the sawdust off with a clean soft cloth, a tack cloth, or the blower side of a vacuum cleaner. If you blow the dust off with compressed air, be very sure there is no condensed moisture in the tank, because from now on, moisture will kill the polish.

HANDLE THE CASE WITH CLOTH OVER YOUR HANDS from now until the oil is cured. Any moisture, even the moisture in your fingertips, will leave obvious marks which will have to be sanded out. The oil will not make those marks go away.

Now is when you find out if you've done your work well. You may find out that the 600 has brought out swirl marks and other scratches you didn't know were there. Those marks came from one of the coarser grades, and you didn't sand well enough with succeeding grades to get them out. Happens all the time. You have to go back to at least 220 (and probably further back than that) and sand your lovely shine away in that place, following the grades back down to 600 again, and the shine will return. Inspect the wood closely (the guests in your living room will); those swirl marks get worse with oil in them.

Three: Oiling

When you have the wood polished to your satisfaction, oil it as soon as you can. Don't let it sit overnight if at all possible. If you can't find Watco in a large paint store, you can get it through some of the catalogs listed in chapter 10. Get two quarts for a large case, and you should have some left over.

The nice thing about oil is that it makes absolutely no difference how you put it on, as long as you get enough of it on. Slap it on with a brush any old way, or spray it on—hell, you can DUMP it on if you want to. Don't pussyfoot around with a damp rag; the object is to saturate the wood, and keep it wet for about half an hour. Some areas will soak up faster than others; re-wet those areas.

Have the case up on its feet when you oil it, and get oil on the whole case as quickly as you can. Sometimes a piece oiled lying down will collect too much oil in the pores, and it'll weep oil for days afterwards. Standing up, the wood can soak up all it can hold without collecting little puddles in the open pores.

After you've kept everything wet for half an hour or so, take a break. Let it stand, out of direct sunlight, for another half hour, or until the oil just begins to thicken a little. If it gets too thick or sticky, a rag dipped in the oil will soften it again. When it's just a little syrupy, wipe off as much as you can with clean soft rags (bedsheets are great) or soft paper towels. You still can't touch it with your hands; use a rag in each hand. Then, with clean dry rags, buff it dry. Set your light up on the other side of the case from you again; it'll show up wet spots and drips. Watch for oil dripping out from under the moldings and other hiding places.

When you are sure the case is as dry and dribble-free as you can get it, go do something else that doesn't kick up sawdust for the rest of the day, and check back now and then to wipe down any oil which may seep out of the pores. You can handle the case (with cloth on your hands) tomorrow.

At this point we find out how much of a fanatic you are. Watco oil dries overnight. However . . . it doesn't actually cure for . . . thirty days. If you could find a clean, warm room to put that case in for thirty days, where little fingers would never touch it, you'd have as close to a perfect finish as you are capable of from then on.

You probably don't want to do that (I've never done it). So do the next best thing. Do not allow the case to come in contact with human hands for thirty days. It's not that hard. Just use clean rags (or cheap cloth gloves) to handle it with when you finish assembly (hanging the doors, setting the locks, etc.) and use rags on your hands when you and a friend move the clock into the house, and then chain off the area, tie a large dog in the living room or sit in a chair with a long stick for the next month, and keep all those admiring fingers away. It's a pain, I know, but you've probably found a lot of things in this project a pain. It's worth it.

One more thing: if for some reason you decide you have to resand something after the oil is applied, you'll have to wait at least thirty days.

Four: Maintaining the Finish

If you have attained a good finish, there should be no reason to ever wax it, but if you ever do decide (after the first thirty days) that you must wax, be absolutely sure to use a dark or colorless paste wax, like Johnson's. NEVER apply a white wax (like Trewax) to this kind of an open pore finish. The white wax will collect in the pores and it will stay there. Looks terrible. Don't wax it. And especially never use spray wax.

Watco puts out a product for maintaining the oil (they also put out a liquid wax). What they (and I) recommend if you like to rub a little something into the wood every six months or so, is called Watco

Satin Oil. It's not the same as the finish you put on in the first place—that was called Watco Danish Oil Finish.

You may have to get the Satin Oil by mail order. You can apply it for the first time when your thirty days are up if you want, or even before. Just rub a little in with a cloth and wipe it off. Good stuff.

8
LEADED GLASS

Something magical happens to that wooden crate you've been perspiring and cursing over all this time when the movement goes in and the leaded glass is installed. Suddenly it's no longer just a big dark box. It comes alive—it becomes a grandfather clock.

Leaded glass is not common in clock cases (I don't know of any manufacturer who uses it), nor is it particularly traditional. But it is so beautiful if it's done right. I've never used anything else. Unlike ordinary window glass, a leaded-glass panel has a high value all its own; it isn't just "one of the boards" of the case.

For one thing, the glass itself is of much higher quality. Stained glass comes in many textures as well as colors, and it comes from many different countries. It's all more interesting than the hardware-store stuff. For the amount of glass you will be using, it would be pretty pointless to compromise by leading up window glass.

You may very likely decide to have your glass done by a professional stained-glass studio, and it's a perfectly justifiable decision. You may reason that you are after all a craftsman in wood; you do that well, and you are not particularly interested in dabbling in such an unfamiliar craft and ending up with what may well be a mediocre result.

There's an investment involved in the tools and time and glass it takes to make even a small clock door panel, and you'd be wise to invest some considerable additional glass and lead in practice before you get down to the real thing.

Depending on the complexity of the design, the amount of art/layout time, the quality of the glass, and the reputation of the studio, you can spend from about $15 to $50 a square foot for a finished panel suitable for a clock. If you elect to do it yourself, your equipment may cost you $50 or so to start, and then you can probably work for $5 to $10 a square foot after that. If you don't think you'll ever use your equipment again, you'll probably be better off if you spend your time doing what you do best, maybe sell a piece of woodwork as a result, and pay a stained-glass person to do what he does best.

A possible alternative which can be a very satisfactory way to have a little of both is to enroll in a class in night school, or through a craft or stained-glass studio. Stained glass is so popular these days that you shouldn't have any difficulty finding a class somewhere.

Tell the instructor when you inquire about the class just what you have in mind; chances are he'll welcome your sense of direction and offer to help as much as he can.

The class will provide most of the equipment, or help you get what you need at a reasonable price, and best of all, you'll probably be able to choose your small pieces from a large assortment of glass, without having to go out and buy large sheets. There will also be a pro right there to help keep you out of trouble.

They'll probably want you to do some little doodad first, a trinket or some other such unclocklike thing, but that's how you learn. Leaded glass requires practice. Who knows? If you stay with it you may just add stained glass to your ever-increasing list of talents.

Whether or not you decide to do the entire job yourself, it's important that you at least be intelligent on the subject, so you can go to the supplier or studio with something in mind and some kind of a design worked up with proper dimensions.

One: Design

Let me emphasize that by "design" I'm not talking about setting a little Tiffany window in the front of your clock case; the glass must be a subtle embellishment, not a separate element which becomes the center of attention. Too much lead and color concentrated in that little panel can make it look like an object stuck in the middle of the clock, like Venus with a barometer in her navel. Interior leadlines should be used with restraint, and they should be the thinnest lead practical—$1/4''$ or even $3/16''$. Remember, you want to see the pendulum and weights; that's why you have glass in the door in the first place.

Here are some workable door panel ideas.

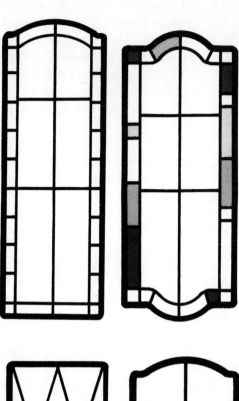

BORDERS . . .
Slightly smaller pieces on top, larger at bottom are more pleasing to the eye.

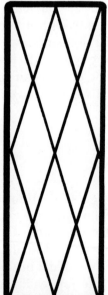

DIAMONDS . . .
Single pale color.

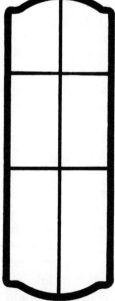

BORDERLESS . . .
Single pale color or colorless.

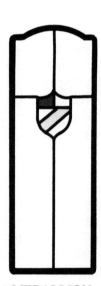

MEDALLION
This could be the trinket you make in stained glass class.

Watch out for difficult or impossible cuts:

This piece would be impossible to cut because of the sharp inside corner.

(It will break here.)

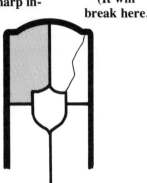

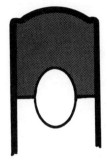

Small radius inside curve is also extremely difficult to cut.

Keep medallions simple. They're too small to contain much lead.

Patterns

First make a pattern for each panel: door front, dial, and side glass. ⅛" masonite is best.

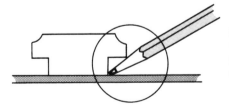

Hold the door solidly against the masonite and with a DULL pencil against the rabbet outline the pattern.

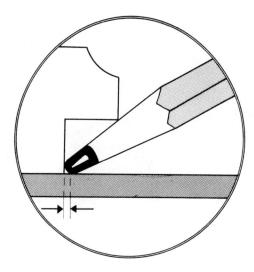

A dull pencil will make a mark a little inside the rabbet. Cutting on this line will give you just the right amount of clearance.

Cut the pattern out and trial fit it. You want to end up with a nice, even, $1/16''$ clearance all around. Sand or cut a little more as necessary. When the pattern is an easy fit but not sloppy, mark it for "up" and "out" so the glass goes in the same way the pattern went in.

Next, on a large piece of tracing vellum (stationery store), draw around the outline of the pattern. This tracing paper will become your "cartoon."

The Cartoon

The outer edge of all your glass panels should be set in ½" or ⅝" lead, whether they are to have any other design or not. If the waist door has a lead edge and the dial and sides do not, the waist door will stand out as an oddball.

For our example let's decide on ½" lead edges.

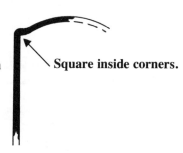

From your outline in, measure off ½" and blacken it in with charcoal or marking pen.

Square inside corners.

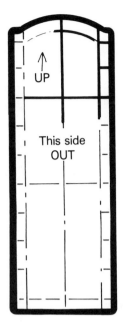

This done, carefully pencil in the interior lead lines and blacken them to ¼" width.

This is your cartoon. Make one for each panel. With these and your masonite patterns in hand, you're ready to see a studio or go to class.

The Cutting Patterns

If you're going to go the rest of the way yourself, you'll need some brown wrapping or butcher paper, some heavy pattern paper, and a couple of large sheets of carbon.

On a smooth, flat surface, tape or pin down a sheet of the pattern paper, then a carbon, next the butcher paper, one more carbon, and finally your cartoon on top. You're going to be making the cutting patterns out of the pattern paper and an assembly pattern out of the butcher paper.

Trace the FULL SIZE outline of the panel, pressing hard enough to transfer through the carbons.

From this line, measure in 9/32" and draw another line all around the perimeter of the panel. This new line is the CUTTING LINE.

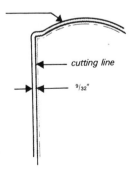

105

THE CUTTING LINE

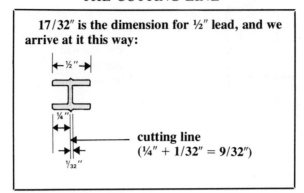

17/32" is the dimension for ½" lead, and we arrive at it this way:

cutting line
(¼" + 1/32" = 9/32")

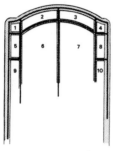

Number each piece.

Draw all the interior cut lines right in the middle of the ¼" leads, and connect them to the perimeter cut line you just measured off. Number the pieces.

Set the butcher paper pattern and cartoon aside, and cut the pattern paper out on the *perimeter cutting line only.*

This will give you a heavy paper pattern exactly 9/32" smaller all the way around than your masonite full-size.

If there are to be no interior cuts (as in the dial glass), then this is your completed cutting pattern.

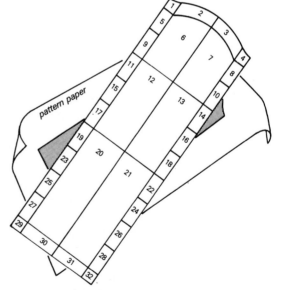

If there are to be interior cuts (waist door), you'll need to buy or borrow (good luck) a pair of pattern shears. Pattern shears are three-bladed scissors. They have two outer blades and a thick center blade which removes a $1/16"$ strip of paper as you cut.

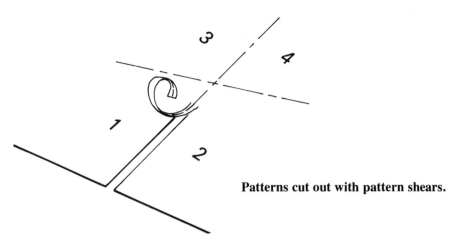

Patterns cut out with pattern shears.

You can see what these dandy scissors do: the $1/16''$ strip they remove automatically makes room between the pattern pieces for the $1/16''$ "heart" of the lead.

Without this clearance between the pieces, the glass would take up all the space in the panel, with no room left for lead.

When you get all the pattern pieces cut out, lay them puzzle fashion on your cartoon.

Two: Choosing Glass

Color

There is one important thing you must understand and remind the studio or supplier about when choosing glass for a clock case: there is little or no sunlight behind the glass in a clock. The usual method of holding a few samples up to a window does not work.

Unfortunately, this fact severely limits our choices. Those bright, beautiful colors you see in the window go absolutely black in a clock case. (Ever been inside a church at night?)

What this leaves us are pale colors, which will usually come out nearly colorless, and the opaque "opalescent" glass of the sort used in Tiffany-type lampshades. But don't despair; remember we're not after a garish, dazzling display in the middle of the case; pale ambers will richen the gold and silver clockworks seen through them, and a little restrained opalescent border will add interest without being shocking.

An even more subtle approach is to use pale transparent colors in the border. Straight on, they appear colorless, but from an angle, when the pendulum or one of the weights is in just the right position, the viewer discovers a touch of color he didn't even know was there.

Types of Glass

The opalescent glass you find will probably be American made. Of the colorless and pale color glass available, the most suitable for clocks is called "antique" or "handblown." Ask for German or French antique. It's not really antique, but it is made by an antique process, involving blowing a cylinder of hot glass, cutting it, and spreading it on a flat, textured table.

The result is a delicate, thin glass with just the right amount of "imperfections" (bubbles and striations) to make it interesting to look at but not hard to see through.

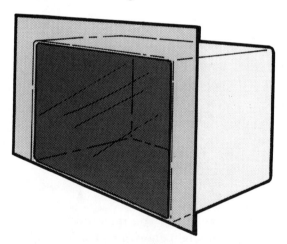

Instead of choosing glass in the usual manner (against a window), simulate a clock case with a cardboard box placed in a darkish corner of the studio. You don't even want direct sunlight on the *front* of the glass, because chances are there won't be any in your living room.

It's a good idea to bring your dial along and place it inside the box for choosing your main colorless or pale glass. (Amber, smoke, and clear look best over brass and silver; greens and blues make the dial look a little sickly.) Prop the dial up about an inch behind the glass and stand across the room and make sure you can read it. Walk around; make sure you can read it from an angle as well.

Three: Cutting

Glass cutters come with or without a ball on the end. I like the ball kind best.

If you have an old one in the shop, it's probably dull. Better buy a couple of new ones.

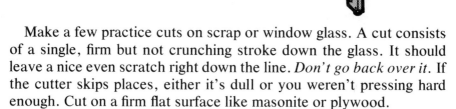

Hold the cutter straight up and down, more or less as you hold a pencil, only tuck the handle between your index and middle fingers.

Make a few practice cuts on scrap or window glass. A cut consists of a single, firm but not crunching stroke down the glass. It should leave a nice even scratch right down the line. *Don't go back over it.* If the cutter skips places, either it's dull or you weren't pressing hard enough. Cut on a firm flat surface like masonite or plywood.

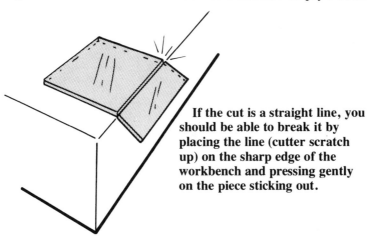

If the cut is a straight line, you should be able to break it by placing the line (cutter scratch up) on the sharp edge of the workbench and pressing gently on the piece sticking out.

If the cut is a curved one, tap gently with the ball directly under the cutter scratch, working from the ends to the middle. You'll be able to see the glass breaking through or hear the tap become hollow-sounding as you complete the break. The two pieces should come apart in your fingers with very little pressure.

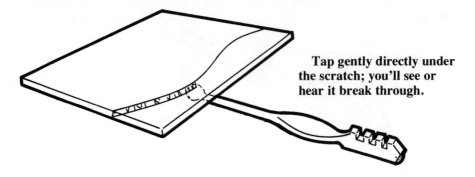

Tap gently directly under the scratch; you'll see or hear it break through.

Practice on scrap until you're fairly confident with the cutter. It shouldn't take too long.

LUBRICATE THE CUTTER

Lubricate the cutterwheel axle frequently by pressing it into cotton dampened with kerosene or light oil.

When you're satisfied you're a pro (or close enough), you can start cutting the real thing. Remember that stained glass has a grain (just in appearance; it doesn't affect cutting). When you lay out your pieces, try not to have the grain going every which way.

The big rectangular pieces are easy. If the edge of the glass sheet is perfectly straight, and you need several rectangles the same width (exactly the same) you can rip off a strip using a straightedge (you might need somebody to hold one end).

With the narrow piece in both hands and a friend holding the sheet down, snap it over the edge of the bench.

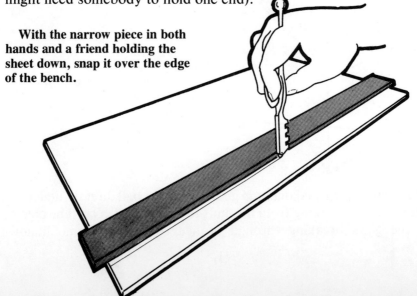

If your rectangles vary a little in width (*believe your patterns*—you can't fudge in leaded glass; $1/16''$ can make a big difference), cut a strip off the edge of the sheet as above, but *at least* $½''$ or better, an inch, oversize. You can't trim thin strips ($¼''$) off the edge of a piece of glass. Not without a lot of difficulty, anyway.

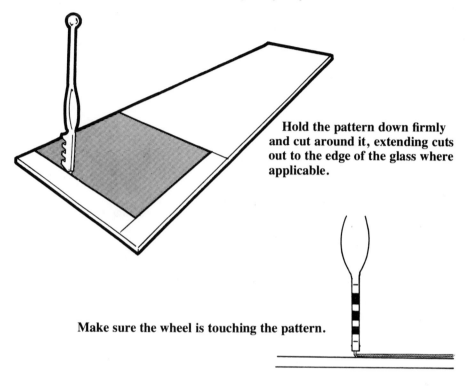

Hold the pattern down firmly and cut around it, extending cuts out to the edge of the glass where applicable.

Make sure the wheel is touching the pattern.

Snap off the big cuts. If the narrow pieces are stubborn, tap them with the ball from underneath.

GROZING

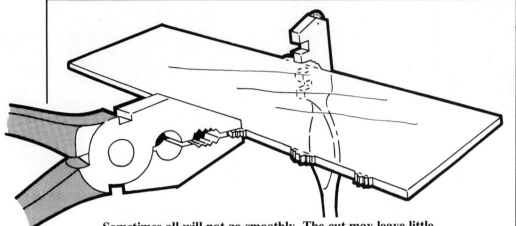

Sometimes all will not go smoothly. The cut may leave little burrs. Grip them with square-nosed pliers and snap them off. This job is also what the teeth on your cutter are for.

If the burrs are very small, crumble them off by rocking the pliers down, causing the jaws to scrape against the sharp edge of the glass like a rasp. This is called "grozing."

For final, very minor touch-ups, glass can be sanded on the belt sander, but don't overheat it.

Always wear safety glasses when cutting and grozing.

Cutting the Dial Glass

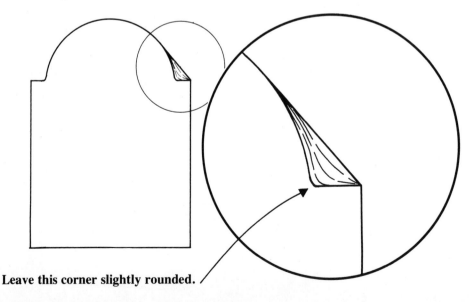

Leave this corner slightly rounded.

If you can, pay the supplier to cut this piece for you. The inside corners are very difficult, and a lot of expensive glass is broken in attempts to cut them. If you hire the supplier to cut it, breakage is his responsibility, not yours. (Chances are, with his experience, he'll pull it off on the first try.) Get a price up front.

If you do have to do it yourself, cut out the panel, bypassing the corners. Then make a series of "relief cuts," tapping them with the ball and breaking them off with pliers one at a time. Be careful.

DEBURRING

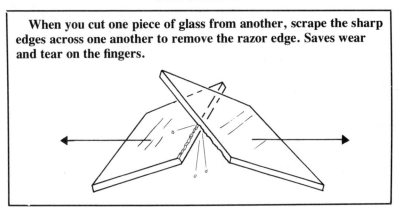

When you cut one piece of glass from another, scrape the sharp edges across one another to remove the razor edge. Saves wear and tear on the fingers.

Four: Glazing

Glazing is "leading up." Lead comes in six-foot lengths called "cames." Determine how much ½" and how much ¼" you will need and buy at least one extra came of each for practice and goofs.

You'll also need the following tools:
a. A rather large soldering iron.
b. 50/50 tin/lead solder with NO core (a small roll).
c. Oleic acid (a tiny amount; maybe a tablespoon) and a flux brush.
d. Horseshoe nails, if you can find them, or 4d box nails (a pound).
e. An inch-and-a-half putty knife to cut down and sharpen into a lead knife (see p. 114).
f. A wooden "lathekin" to open up the lead (see below).
g. A couple of strips of wood, about ⅜"×¾".

LEAD KNIFE

Cut the blade in half with a hacksaw.

Sharpen until it is slightly curved and very sharp (see chapter 2).

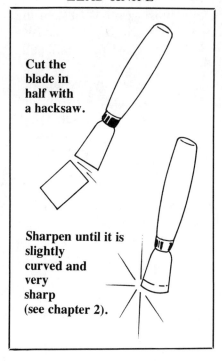

LATHEKIN

From ¾" hardwood, rip a 5/16" strip about 6" long.

Taper one corner to about ⅛" and round all edges

SMALL CORNER opens lead for glass.

FAT CORNER opens came end for joints.

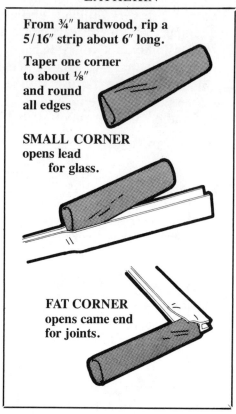

Lay the butcher paper assembly pattern down in front of you on a smooth plywood surface. Along the front edge, exactly on the full-size line, nail one of the wood strips. If your door glass has a straight top or bottom, nail another strip on this full-size line 90 degrees from the first one.

Place the stops on the full size line and nail them to the workbench.

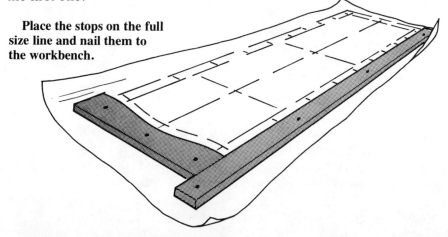

If the top and/or bottom are curved, it may be helpful to transfer the full-size curve from your masonite pattern to a piece of plywood and cut it out for an end stop.

PULLING THE LEAD

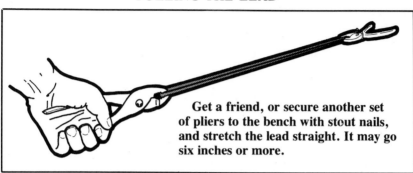

Get a friend, or secure another set of pliers to the bench with stout nails, and stretch the lead straight. It may go six inches or more.

Start by cutting a ½" lead the exact length of the full size of the side. Open the inside edge of this piece with the lathekin and tuck another ½" lead into the end, forming the first corner. You will build the panel from this corner out.

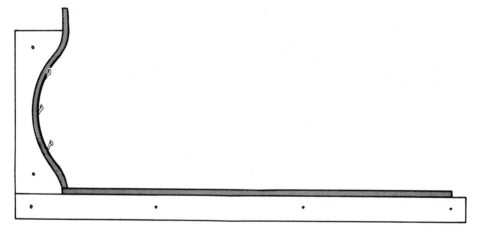

Bend the second strip to fit inside the curved wooden stop and temporarily hold it in place with nails (horseshoe nails are best, because they are flat). Leave this piece extra-long for the time being.

Set your first piece of glass, tapping it snug with a hammer handle. Tuck the first piece of ¼″ interior lead into the ½″ perimeter lead, and cut it off just a little ($^1/_{32}$″) shorter than its corresponding glass edge.

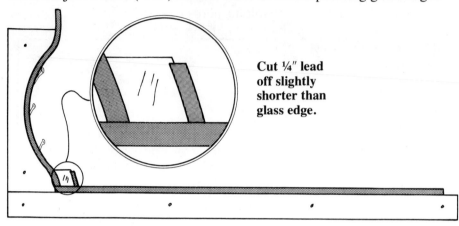

Cut ¼″ lead off slightly shorter than glass edge.

This done, tuck a long piece of ¼″ lead into the perimeter as shown, inserting the short piece into it. Hold this interior lead snugly against the first piece of glass with nails.

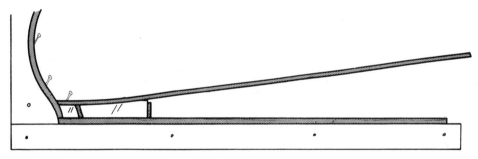

Insert the remaining border glass pieces and their "end" leads the length of that side, pressing the long piece down and tapping in nails as you go. As you insert each piece of glass, tap it in place with hammer handle.

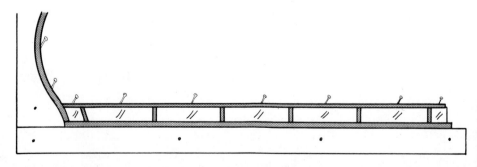

With any luck, you should end up with a neat border along one side. DON'T SOLDER ANYTHING YET. That has to wait until the entire panel is assembled to your satisfaction.

It's now just a matter of building across to the other side and adding or removing nails as necessary, tapping each piece of glass in place until it fits precisely against the cut line of the assembly drawing underneath.

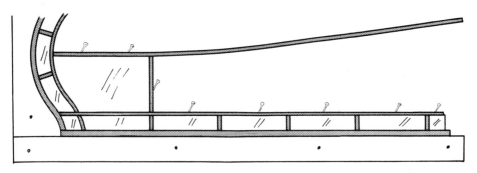

If for some reason you decide to butt the lead joints instead of tucking one inside the other, it will be necessary to cut each piece precisely to length as below.

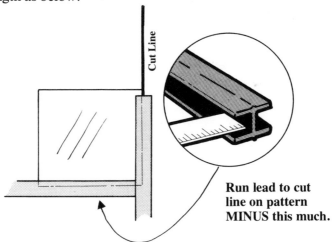

If you were careful to tap each piece into its precise place on the template, you should have a glazed-up panel just the correct full size to fit your door. If it turns out a bit oversize (it's not very likely to come out undersize), cut two more pieces of wood, one for the side and one for the end, and tap them with the hammer against the open side and end until the panel compresses to the correct size. Go easy; it'll compress some, but don't get carried away. This is not exactly a "beat to fit" operation.

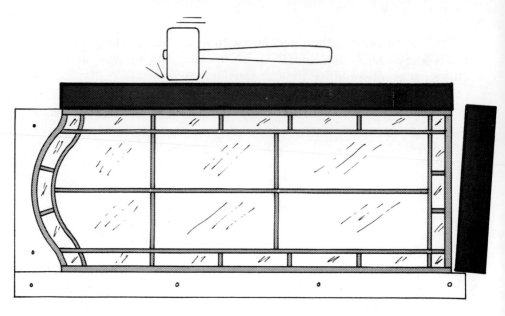

When the panel is all soldered up, you'll pinch the outer leaves of the perimeter lead together and this will reduce overall dimensions a fraction, if that helps.

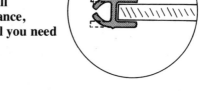

Closing perimeter lead gains a small amount of clearance, which may be all you need for a good fit.

Five: Soldering

Time out for a crash course in soldering before you attack your hard-earned panel.

Before you start, plug in the iron and let it get as hot as it's going to get, file the point clean, and "tin" it with a little solder.

Cut several short lengths of lead and assemble some sample "T" and "L" joints on the workbench. Dab them all with a small "flux brush" dipped in oleic acid.

When you're sure the iron is hot, expend your first two or three sample joints by holding the iron on them too long and melting them. This will give you a feel for how long it takes (and it's not all that long) to overheat a joint.

With the rest of the joints, practice. Learn how much solder to add to make a smooth joint without dribbling it all over the place, and learn how long to hold the iron on the joint so the solder flows neatly but the lead cames don't melt.

An interesting point in our favor: the melting point of tin/lead alloy (solder) is lower than either tin or lead separately. This means that solder *will* melt before the lead cames every time; you just have to know when to stop. Some stained-glass folks plug the iron into a heavy-duty rheostat, set to allow the iron to get hot enough to melt solder but not hot enough to melt pure lead. Effective, but expensive and not entirely foolproof.

Flatten the joints with the end of the hammer handle and paint them with oleic acid. Wipe off any excess with a rag and with a little roll of coreless solder in one hand and the iron in the other, commence.

Touch the end of the solder to the joint, and put the iron on top of the solder. In just a second the solder should melt and the tip of the iron will be touching the joint. Hold it there only until the solder flows neatly onto and into the joint. Feed a little more solder if necessary. The second you're done, get the iron off.

If you should melt through the lead, it is possible to bridge the gap with solder, but by the time you're good enough to do that you probably won't be melting through in the first place.

Don't feel bad if it takes you all day or all week to get enough confidence with the iron to start soldering up your panel; the people who do it for a living have been practicing for years.

When you do get your door panel all soldered up on one side, it's still pretty fragile, and there's a trick to turning it over. Remove the wooden stops and slide the panel out over the edge. Supporting it with your hands spread wide, pivot it down to vertical. Now you're holding this thing vertically against you, and the problem is to get the side next to your stomach, the soldered side, turned around and facing the workbench, so it can be pivoted back up. Easiest thing is to hand the panel to a friend (make sure it's a good friend) and have him or her put it back on the bench.

Then solder the back side.

Pivot the panel down over the edge, turn it around, and pivot it back up. Stained glass panels should always be carried and stored vertically, on edge.

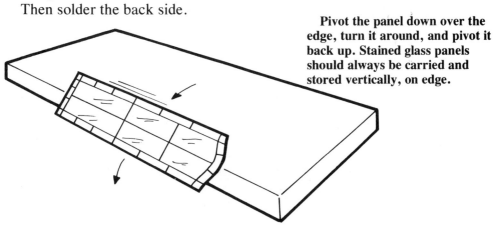

Six: Cementing

Cementing is technically for waterproofing stained-glass windows, but it does a lot more than waterproof. The plaster in the cement compound stiffens and strengthens the panel, and the lampblack dulls the lead to a dark antique patina.

Cementing often makes the difference between professional stained glass and soldered-up trinkets.

It's a messy job, so get some rubber gloves and spread plastic over the workbench.

Aside from the cement materials and a bucket or gallon can to mix them in, you'll need a pointed stick (pencil size), a heavy wooden spoon or paddle, and a wooden scrubbing brush sawed in half to make two.

Cut a scrubbing brush in half; one half will be for cementing, one for cleaning.

The Recipe*

½ c boiled linseed oil
3 oz. lampblack
½ c Japan drier
⅛ c turpentine
1½ c whitening (pure plaster)
⅛ c plaster of paris

The mixture should be quite thick, but not too stiff to stir. Use plaster of paris to thicken; turpentine to thin.

*courtesy of The Judson Studios, Los Angeles, Calif.

Lay the panel flat on the bench and plop a big spoonful of the cement right in the middle. Take one of the scrubbing brushes and smear the goop all over the panel, being especially careful to brush it under the leaf of all the leads. Add more cement as needed.

When all the lead is darkened and cemented, remove as much excess as you can with the same scrubbrush and scrape it back into the bucket.

It's best to let it set a couple of hours before you turn it over and cement the other side. An hour or two after the second side is cemented you can clean the panel.

Cleaning

Clean up against the lead leaf with the pointed stick. Make sure you get the excess out of the corners. Then sprinkle half a cup or so of plaster of paris on the panel and scrub with the clean dry scrubbrush.

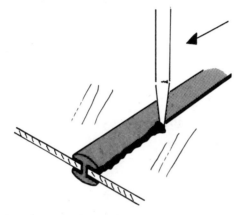

Dry plaster will clean that nasty black cement up slicker than you can believe and, brushed under the lead leaf, it will help dry the cement under there. Blow the excess plaster off with compressed air, or sweep with a soft bench brush or paintbrush and polish the glass with a cloth.

Properly cleaned, the panel should have cement only under the lead, and the leads should be stained dark gray or black.

Your leaded panel is finished; put it aside, at least for the night, before installing it.

Seven: Installing

Some of the usual techniques for installing glass don't work very well in clocks; in a ⅜" rabbet there isn't really room for wooden strips or quarter round, and glaziers' "push points" are difficult to get into hard wood.

Here are two ways which work:

The wooden peg method is my favorite, and just between us, the best peg for the job is the bottom of a common kitchen matchstick. It's the perfect size for a snug fit in a ⅛" hole.

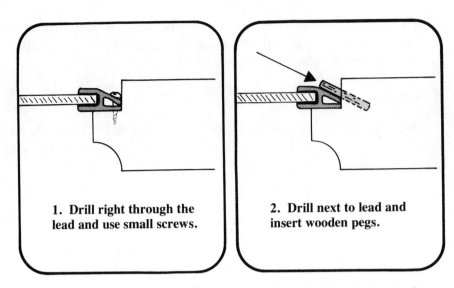

1. Drill right through the lead and use small screws.

2. Drill next to lead and insert wooden pegs.

Cut the pegs about ¾" long and push them in ⅜" deep holes *without glue,* about every six inches or so around the perimeter of the leaded glass.

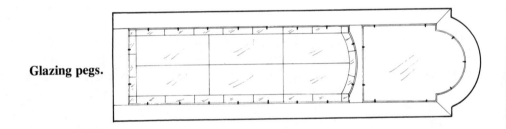

Glazing pegs.

If the glass ever needs repair, the unglued pegs come right out with pliers.

9

WOOD CARVING

Carving is one of the most satisfying ways of decorating and individualizing furniture. Clocks lend themselves well to carving, because of the flat surfaces in the base, the crown, and even the door, just waiting for some kind of decoration.

If at all possible, the carving should be done before assembly.

One: Tools

You don't have to be a sculptor to do surface decoration or "relief" carving; you just have to have a little time, and four or five sharp, well-chosen cutting tools. And if you have a router, you can even cut the time significantly. But the first thing you need to equip yourself with is a good solid carving bench. It won't cost much, and it'd pay you to stop and build one.

The Bench

The most important thing in a wood-carving bench is MASS. If you want your downward cuts ("stop cuts") to sink into the wood, the workpiece has to be backed up with what is effectively an immovable object. No matter how sturdy the legs of your table are, if the top gives or springs at all during the moment of impact, the blow is dampened, and the cut is ineffective.

The bench top should be a thick, heavy object which backs up the cut with its own inertia—like cracking nuts on a rock.

Cutting Tools

The easiest way to get educated about chisels, gouges, and the like is to order the mail-order catalogs from Woodcraft and Garrett Wade Co. (see chapter 10). These are both handsome catalogs. They cost about a dollar apiece, and they are as informative as they are beautiful.

WOOD-CARVING BENCH

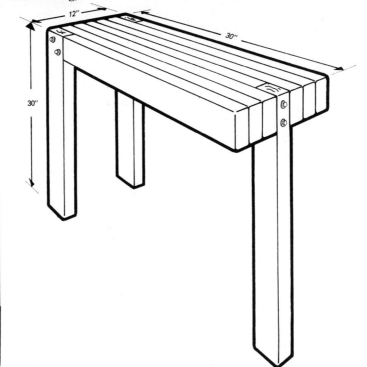

Glue up some construction two-by-fours to make a little bench like this one. Leave space to bolt in three legs (three legs make it self-leveling).

It'll only cost you three eight-foot two-by-fours, and even when you're not carving you'll find it handy to have a solid place to beat on things once in a while.

The top doesn't have to be pretty, but it should be reasonably flat and true.

The tools these places sell are of as high quality as you're going to find anywhere, and Garrett Wade also carries a line of those gorgeous European workbenches.

For surface decoration carving you're only going to need a few carefully selected tools, so by all means *do not buy a set*. Carving sets are for tinkerers who don't have a specific carving direction, and you'll likely end up with some stump-carving tools you'll never use. These things cost $6 to $10 a blade, so choose very carefully.

Briefly, the terminology is as follows:

CHISEL: just like a carpenter's chisel. In fact they can *be* carpenter's chisels if you already have some good ones.

SKEW: a chisel with an angled cutting edge. Useful for getting into tight places.

GOUGE: shaped like lathe gouges, with a curved cutting edge (lathe tools are too big for wood carving).

PARTING TOOL: V-shaped cutting edge for cutting V-shaped grooves.

VEINER: tiny U shaped gouge for making—you guessed it—tiny U-shaped grooves. (Who said wood carving was complicated?)

There are more specialized tools, but these are all you should need.

You'll also want to buy or turn a round "potato masher" sculptor's mallet. It's round so the striking face of the mallet will always be reasonably parallel to the chisel butt; you don't have to concentrate on keeping the hammer head in exactly the correct position. The mallet you buy will probably be oak or lignum vitae; lignum vitae is good because it is dense and heavy (like your bench, it doesn't "give"). You might buy a lignum vitae "slammer" about 3″ across the big end, and then turn another out of oak, 2¼″ to 2½″ across for more delicate work.

So that it'll mean something to you, I'll recommend specific cutting tools as we proceed through the fundamentals of carving.

We also ought to get into the fundamentals before we talk about design, so you'll know what your design limitations are.

Two: the Cuts

The design is a little picture—an object or group of objects—which is left standing while all the space around them, the background, is "lowered."

Stop Cut

The first cut is the stop cut. This is the one that requires the solid bench, the solid mallet, and a little violence. The stop cut outlines the object, and it is cut straight down into the wood all the way around the perimeter of the design, about ⅛″ deep.

Theoretically, you should own a gouge with the same curve in its cutting edge as every curve in your design. This very quickly gets ridiculous, as do your attempts to explain to your spouse how for that

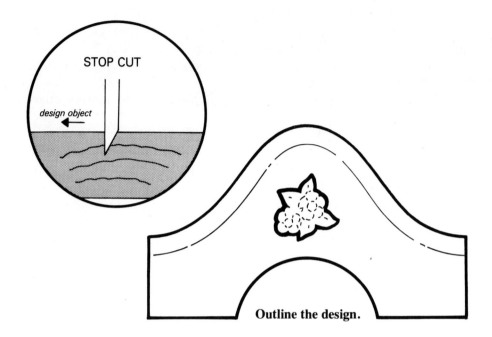

Outline the design.

one whack at that one little curve you had to buy that ten-dollar gouge.

If you're a serious wood carver, you probably will buy a new gouge every time you need one, and in a few years you'll have three hundred of them and you're in business.

A slightly more conservative approach, which is nearly as effective, is to buy one very narrow *straight* chisel—3/16" to 1/4" across. Such a narrow chisel will follow the contour of most curves by making a series of tiny straight cuts:

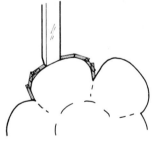

Make certain the stop cut completely outlines the object in an unbroken line to an even depth of about 1/8".

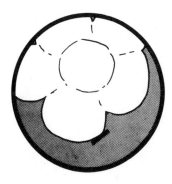 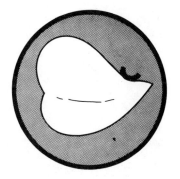

In the case of very small outside curves, the overhanging "corners" of the cut will be in the waste, and therefore no problem.

If the cut is a small *inside* curve, shift to a small gouge. The corners of the tool will again cut into the waste and not the design.

Lowering

For this cut you'll want a couple of fairly "slow" (shallow) gouges; look for a #5 gouge, one about ½" wide, and another about 5/16". (Number indicates *depth* of curve, dimension indicates *width* of blade.)

Clamp the workpiece to the bench, or nail wooden stops in two places so the pieces can be worked against them. You need to be able to rotate the piece, but you don't want it to move by itself.

Try a couple of wooden discs for stops. Nail them to the bench.

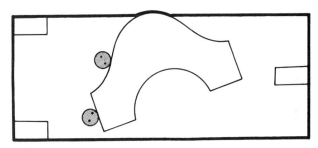

Work with the slow gouge all around the "stopped" design until it is outlined with a "lowering cut" at least ⅜" across.

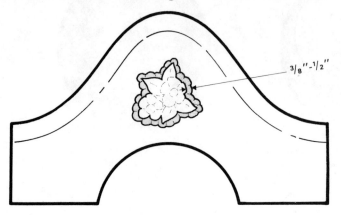

Once the object is cleanly outlined, you can leave the background the way it is, or you can lower all the way out to the edges of your workpiece (almost).

If you decide to lower the whole piece, you'll have to make another stop cut around the outside where you want the lowering to end.

NOTE:

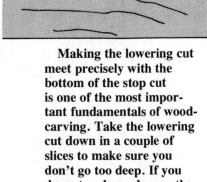

Making the lowering cut meet precisely with the bottom of the stop cut is one of the most important fundamentals of woodcarving. Take the lowering cut down in a couple of slices to make sure you don't go too deep. If you do go too deep, deepen the stop cut until the two meet.

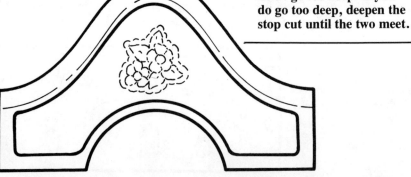

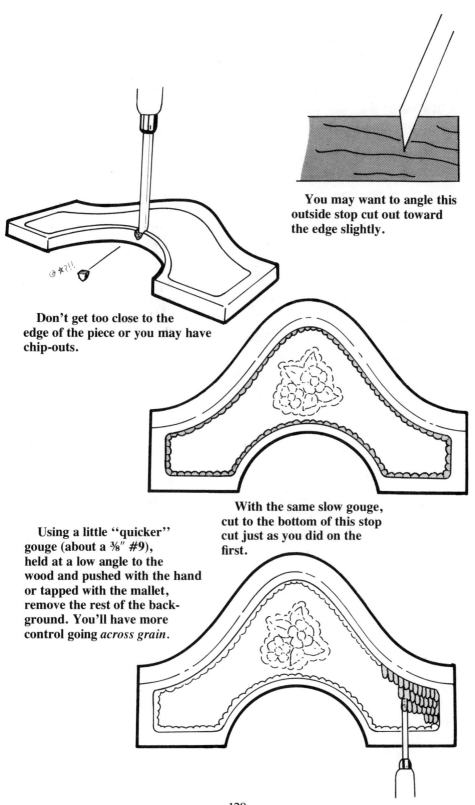

You may want to angle this outside stop cut out toward the edge slightly.

Don't get too close to the edge of the piece or you may have chip-outs.

With the same slow gouge, cut to the bottom of this stop cut just as you did on the first.

Using a little "quicker" gouge (about a ⅜" #9), held at a low angle to the wood and pushed with the hand or tapped with the mallet, remove the rest of the background. You'll have more control going *across grain*.

As the name implies, a "quick" or deep gouge removes wood quickly. Unfortunately it leaves the background with more texture than it really should have. To make it more subtle, go over the background again, this time with the "slow" ½" #5 gouge. The background will still look and feel handcarved, but it won't be as distracting.

THE RIGHT (AND WRONG) DIRECTION

Look at the grain in the edge of the workpiece. Always work in the "uphill" direction, just as you would with a plane, a shave, a jointer or planer, or any slicing tool.

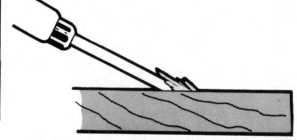

Working in the wrong direction, even the sharpest blade will run down the grain instead of slicing it off. Splintering results.

Modeling

With the background all lowered away, we can go to work on the design objects. "Modeling" is our attempt to make a flower look somewhat flowerlike, or a leaf somewhat leaflike, by cutting some parts deeper than others, rounding things where appropriate, and so on.

There's nothing mysterious about it; it involves the same stop and lowering cuts we used in the background.

A SIMPLE FLOWER

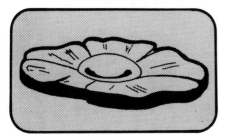

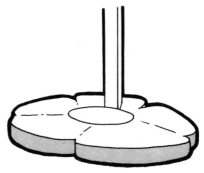

1. Outline center with a stop cut.

2. With the slow gouge, scoop out the petals, tapering them from the edges down to the bottom of the stop cut around the center.

3. Vary the surface of the petals; make the edge of one a little higher than the one next to it.

4. With the slow gouge upside down, round the center.

The finished flower should look something like this in cross section.

THE RIGHT DIRECTION (AGAIN)

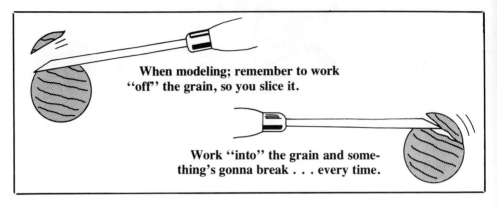

A SIMPLE LEAF

1. In nature, leaves tend to twist around a stiff center vein. To us, this means that the raised parts (U) and the lowered parts (D) are on opposite "corners." Mark them on the entire design with a pencil before you start cutting.

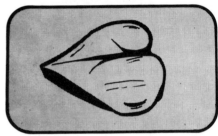

2. Stop-cut down the center, getting deeper as you go, from the point down to the base.

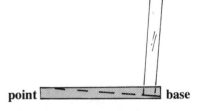

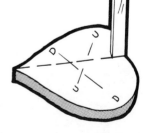

Where one leaf overlaps another, raise that corner of the top leaf and lower that corner of the bottom leaf.

Just for illustration, look at the leaf as a rectangle. Notice that the "up" corners slide down to the bottom of the center stop cut, while the "down" corners form a shoulder at the center, and taper down to the outside.

3. When you're finished and satisfied with the modeling (you can sand or steelwool a little if necessary, but this is a no-no with hard-core woodcarvers, so don't tell), take a small parting tool and cut a couple of small "veins."

GRAPES

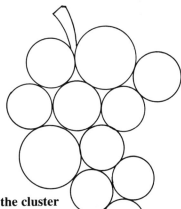

1. Don't make the cluster too symmetrical, and vary the size of the grapes a little; from dime size to about quarter size.

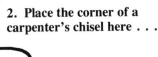

2. Place the corner of a carpenter's chisel here . . .

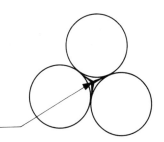

. . . and make three tapered stop cuts.

133

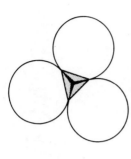

3. With the same chisel, remembering to slice "off" the grain, connect the points of the three-pointed star, forming a little inverted pyramid . . .

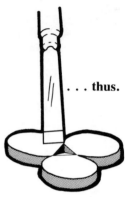

. . . thus.

Make these cuts every place three grapes meet, and make them *clean* (no fuzzies at the bottom).

With the slow gouge upside down, round the grapes.

ACORNS

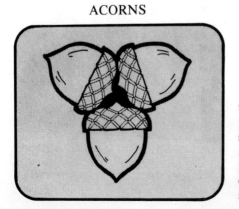

By now, you should be getting the hang of it, and acorns should present no real problems. Just stop-cut around them, model them, and crosshatch the caps with the parting tool. The only thing different about them is the shallow stop-cut where the nut meets the cap.

Any of these objects (or any others) you choose should obviously be well rehearsed on scrap before you tackle the clock case, and even the final piece should be carved before it is installed in the case.

Three: Using a Router

If you are carving a large area and you have a lot of background lowering to do, there's no reason not to let the machine take care of some of the drudgery for you. The end result will be the same or better.

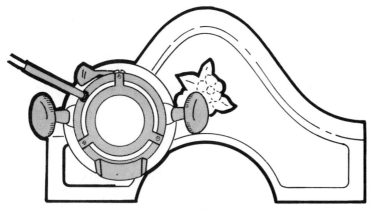

1. With the piece clamped down and your goggles on, sit low enough to see where the bit enters the wood, and make your "stop cuts" with a ⅛" straight bit.

Cut ⅛" to ¼" deep. Work very carefully and this can be virtually your finished stop cut.

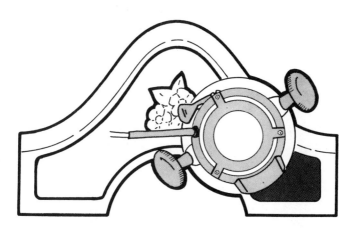

2. Insert a ⅜" straight bit and, working from one side to the other, remove (lower) the entire background.

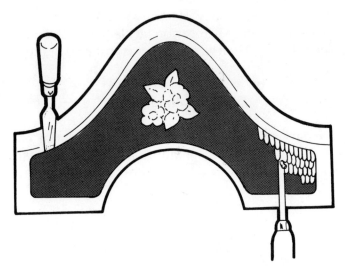

3. With chisels, dress up the stop-cut where necessary, and texture the background with the slow gouge.

You just saved yourself *hours*—and nobody will ever know.

Four: Design

Just like anything else, wood carving demands that you know your limitations. Work up a design you know you can be successful with. If you practice the techniques I've talked about you won't be doing *The Last Supper,* but you'll make a heck of a nice nondescript-daisylike-all-purpose flower, and a basic leaf, not to mention grapes and acorns (although not all on the same clock, I hope).

More specific and botanically correct flowers and leaves are just a variation away from my basic examples, but you don't have to get too scientific; just try to use oaklike leaves with acorns and grapelike leaves with grapes, etc.

Following are some idea sketches to get you started.

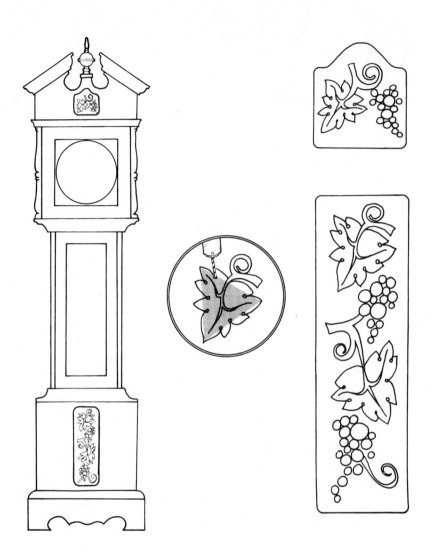

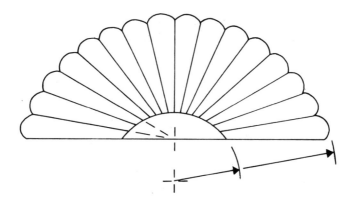

SIDES

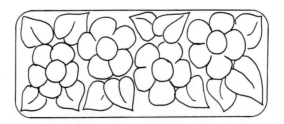

10

SOURCES

Following are two lists of mail-order supply houses who cater to craftsmen. You're probably familiar with some of them; most of them advertise in the major magazines. The lists are organized in a very loose subjective order, more or less corresponding to my own personal preference (which naturally may have little resemblance to YOUR personal preference).

Each catalog has some things the others don't have, and there is an education to be had in just reading them, so the more serious you are about woodworking, the more catalogs you should obtain.

An asterisk (*) indicates that I have done business with them.

Clock Movements, Dials, Plans, Kits, and Parts

1. Craft Products Co.*
St. Charles, Ill. 60174
Catalog $1.50 116 pp
Probably the largest selection of movements and dials. They also have plans and a few case kits, clock hardware, and books. I use this company quite a bit.

2. H. DeCovnick & Son*
Box 68
Alamo, Calif. 94507
41 pp catalog $1.00
Good west coast supplier, lots of movements, dials, and nice plans and kits.

3. Kuempel Chime Clock Works and Studio*
21195 Minnetonka Blvd.
Excelsior, Minn. 55331
Free Brochure

Their movement/dials come in sets. Also very nice plans and kits.

4. Mason & Sullivan Co.*
39 Blossom Ave.
Osterville, Mass. 02655
31 pp catalog $1.00
On Cape Cod. Nice little color catalog, limited but balanced selection of movements, dials, plans, kits, and clock hardware.

5. Newport Enterprises
2313 West Burbank Blvd.
Burbank, Calif. 91506
60 pp catalog $1.00 (refundable)
Movements, including several battery and plug-in electrics, dials, including round dials and paper dials, and numerals for making your own dials. Some hardware.

6. Emperor Clock Co.
Emperor Industrial Park
Fairhope, Ala. 36532
Free brochure
They bill themselves as the "World's Largest Manufacturer of Grandfather Clocks." Kits, movement/dial sets.

7. Viking Clock Co.
The Viking Building
Foley, Ala. 36536
Free brochure
Kits, a couple of movement/dials.

Goodies (Hardware, tools, wood, veneer, finishes, moldings, etc.)

1. Craftsman Wood Service*
1735 Cortland Ct.
Addison, Ill. 60101
152 pp catalog 50¢
No craftsman should be without this and the next few catalogs. The cover of this one says "Over 3500 Items." Very handy. Tools, wood, veneers, etc.

2. Constantine*
2050 Eastchester Road
Bronx, N.Y. 10461
100 pp catalog 50¢
Like Craftsman Wood, Constantine is a good basic supplier of all the nutsy-boltsy stuff you can't usually find in your local hardware store. They also carry books and a good selection of veneering/marquetry supplies.

3. The Woodworker's Store*
Industrial Boulevard
Rogers, Minn. 55374
100 pp catalog $1.00
Similar to nos. 1 and 2, but of course each catalog is likely to have something the others don't.

4. Woodcraft Supply Corp.*
313 Montvale Ave.
Woburn, Mass. 01801
96 pp catalog $1.00
The Woodcraft and Garrett Wade (no. 5) catalogs are a joy to own. Woodcraft is an extremely quality-conscious, craftsman-oriented supplier of tools, particularly wood-carving tools. This is also the place to buy wooden planes, imported cabinet chisels, Japanese pull saws, adzes—you name it. They also do a little article on a working craftsman on the cover of each issue. I like them a lot and I don't mind saying so.

5. Garrett Wade*
161 Avenue of the Americas
New York, N.Y. 10013
100 pp catalog $1.00 — Inca power tool catalog $1.00
The pages of these catalogs are suitable for framing. Please buy something from them so they can pay their photographer. As with Woodcraft (no. 4), everything Garret Wade carries is beautiful and imported and designed to last about a thousand years. Their special thing is the Old World Workbench. These catalogs contain a lot of good information as well as good tools.

6. Brookstone Company*
126 Vose Farm Road
Peterborough, N.H. 03458
68 pp catalog
The "Hard-to-find tools" catalog. Not strictly for woodworkers, Brookstone has lots of interesting things—tools, gifts, and gadgets.

7. Gaston Wood Finishes*
Box 1246
Bloomington, Ind. 47401
41 pp catalog
Stains, finishes, hardware, some wall clock kits.

8. Bob Morgan Woodworking Supplies
1123 Bardstown Rd.
Louisville, Ky. 40204
15 pp catalog free
Veneering/marquetry supplies. A small catalog which includes three pages of veneering instructions.

Square Nails

Tremont Nail Company
Box 111
Wareham, Mass. 02571
Free Catalog

Spring Miter Clamp

Woodcraft "goodies" no. 4 above

Steel 45/90 Draftsman's Square

Woodcraft, "goodies" no. 4 above